Rotary Wing Aerodynamics

Rotary Wing Aerodynamics

Editor

Alex Zanotti

MDPI • Basel • Beijing • Wuhan • Barcelona • Belgrade • Manchester • Tokyo • Cluj • Tianjin

Editor
Alex Zanotti
Department of Aerospace
Science and Technology
Politecnico di Milano
Milano
Italy

Editorial Office
MDPI
St. Alban-Anlage 66
4052 Basel, Switzerland

This is a reprint of articles from the Special Issue published online in the open access journal *Energies* (ISSN 1996-1073) (available at: www.mdpi.com/journal/energies/special_issues/Rotary_Wing_Aerodynamics).

For citation purposes, cite each article independently as indicated on the article page online and as indicated below:

LastName, A.A.; LastName, B.B.; LastName, C.C. Article Title. *Journal Name* **Year**, *Volume Number*, Page Range.

ISBN 978-3-0365-3576-0 (Hbk)
ISBN 978-3-0365-3575-3 (PDF)

Cover image courtesy of Alex Zanotti

© 2022 by the authors. Articles in this book are Open Access and distributed under the Creative Commons Attribution (CC BY) license, which allows users to download, copy and build upon published articles, as long as the author and publisher are properly credited, which ensures maximum dissemination and a wider impact of our publications.
The book as a whole is distributed by MDPI under the terms and conditions of the Creative Commons license CC BY-NC-ND.

Contents

About the Editor vii

Preface to "Rotary Wing Aerodynamics" ix

Alex Zanotti
Rotary Wing Aerodynamics
Reprinted from: *Energies* 2022, 15, 2072, doi:10.3390/en15062072 1

Riccardo Piccinini, Matteo Tugnoli and Alex Zanotti
Numerical Investigation of the Rotor-Rotor Aerodynamic Interaction for eVTOL Aircraft Configurations
Reprinted from: *Energies* 2020, 13, 5995, doi:10.3390/en13225995 7

Alberto Savino, Alessandro Cocco, Alex Zanotti, Matteo Tugnoli, Pierangelo Masarati and Vincenzo Muscarello
Coupling Mid-Fidelity Aerodynamics and Multibody Dynamics for the Aeroelastic Analysis of Rotary-Wing Vehicles
Reprinted from: *Energies* 2021, 14, 6979, doi:10.3390/en14216979 35

Fabrizio De Gregorio, Antonio Visingardi and Gaetano Iuso
An Experimental-Numerical Investigation of the Wake Structure of a Hovering Rotor by PIV Combined with a Γ_2 Vortex Detection Criterion
Reprinted from: *Energies* 2021, 14, 2613, doi:10.3390/en14092613 63

Sara Muggiasca, Federico Taruffi, Alessandro Fontanella, Simone Di Carlo and Marco Belloli
Aerodynamic and Structural Strategies for the Rotor Design of a Wind Turbine Scaled Model
Reprinted from: *Energies* 2021, 14, 2119, doi:10.3390/en14082119 83

Mohammad Hassan Ranjbar, Behnam Rafiei, Seyyed Abolfazl Nasrazadani, Kobra Gharali, Madjid Soltani, Armughan Al-Haq and Jatin Nathwani
Power Enhancement of a Vertical Axis Wind Turbine Equipped with an Improved Duct
Reprinted from: *Energies* 2021, 14, 5780, doi:10.3390/ en14185780 105

Alexander Štrbac, Daniel Heinrich Greiwe, Frauke Hoffmann, Marion Cormier and Thorsten Lutz
Piloted Simulation of the Rotorcraft Wind Turbine Wake Interaction during Hover and Transit Flights
Reprinted from: *Energies* 2022, 15, 1790, doi:10.3390/en15051790 121

About the Editor

Alex Zanotti

Alex Zanotti, Ph.D., has been an Assistant Professor of Fluid Dynamics at Politecnico di Milano since 2016. He is the Scientific Coordinator of the Experimental Aerodynamics Laboratory of the Department of Aerospace Science and Technology of Politecnico di Milano. He obtained his M.Sc degree in Aerospace Engineering in 2006 and Ph.D in Rotary Wing Aircrafts in 2012 from Politecnico di Milano. He is an aerodynamics specialist and his main research interest is rotorcraft aerodynamics with a particular interest in retreating blade dynamics stall control and blade vortex interactions. In recent years, he has been involved in aerodynamics and aeroacoustics investigations of innovative configurations for Urban Air Mobility vehicles such as eVTOL aircraft. He is responsible for the mid-fidelity aerodynamic solver DUST. He was involved in several EU-funded projects (FP6 and FP7, CleanSky and CleanSky2 programmes) regarding helicopter aerodynamics and tiltrotor aerodynamics. He is in charge of the chair of Fundamentals of Aerospace Engineering course at Bachelor Degree in Aerospace Engineering of Politecnico di Milano.

Preface to "Rotary Wing Aerodynamics"

Rotary wing aerodynamics have historically been widely investigated due to the large number of applications of this discipline in several fields of engineering. A deep knowledge of the main phenomena related to rotary wing aerodynamics such as dynamic stall or blade vortex interactions (BVI) is essential for the design of novel VTOL aircraft configurations as tiltrotors or compounds. In recent years, the great interest and development efforts devoted to new designs of unconventional urban air mobility VTOL aircraft based on electric distributed propulsion have focused on rotor–rotor and rotor–body aerodynamic interactions. A deeper insight into these complex aerodynamic interactions is required for the optimization of the design process for novel aircraft configurations as they affect their performance, structural dynamics, handling qualities and acoustic impact. Moreover, the investigation of the main issues of rotary wing aerodynamics is essential in the field of wind energy for the development of novel wind turbine concepts or for the design of wind farms.

This book contains the articles included in the *Energies* Special Issue "Rotary Wing Aerodynamics" which aimed to collect state-of-the-art experimental and numerical studies showing the most recent advancements in the field of rotary wing aerodynamics and aeroelasticity. I would like to thank all the authors for the precious work devoted to the activities disseminated in the present collection.

Alex Zanotti
Editor

Editorial

Rotary Wing Aerodynamics

Alex Zanotti

Dipartimento di Scienze e Tecnologie Aerospaziali, Politecnico di Milano, Via La Masa 34, 20156 Milan, Italy; alex.zanotti@polimi.it

1. Introduction

Rotary wing aerodynamics represents a widely investigated topic due to this discipline's large number of applications in several fields of engineering and physics. Indeed, rotating lifting bodies provide quite complex and unsteady flow structures that have a robust influence in rotorcraft, aeronautical propulsion, turbomachinery and wind energy fields. Consequently, a deep knowledge of the main classical phenomena related to rotary wing aerodynamics, such as dynamic stall or blade–vortex interactions (BVI), to cite a few, is an essential step to improving the performance of helicopters or wind turbines.

In recent years, research effort in the field of rotary wing aerodynamics was focused on the study of rotor–rotor and rotor–body aerodynamic interactions. This interest was influenced in the aeronautical field by the recent great development efforts devoted to the design of unconventional vertical take-off and landing (VTOL) aircraft for urban air mobility (UAM). Indeed, recent improvements in electric motors and battery technologies present an opportunity for new concepts of personal aviation that will provide benefits to ground traffic in overcrowded metropolitan areas and will also improve the performance of logistics services. Distributed electric propulsion represents a key feature in the design of these new VTOL air vehicles, well-known as eVTOLs. Their architecture is characterised by multi-rotor and multi wing configurations that highlight unprecedented aerodynamics challenges with respect to classical aircraft or rotorcraft configurations. Indeed, the occurrence of several different interactional effects between propellers and lifting bodies has a profound impact on aircraft performance and noise impact. Thus, a deeper understanding of the complex interactional aerodynamics features characterising eVTOL vehicles represents a milestone to be achieved before the next-generation UAM aircraft can soar through the skies of our metropolitan areas. In recent years, the field of wind energy research has also paid great attention to the phenomena of rotor–rotor interactional aerodynamics due to the great effort spent on the development of wind farms. Indeed, a thorough understanding of the complex aerodynamic interactions occurring between wind turbine wakes or the study of effective wake redirection techniques can be considered essential key points to improve power capture and reduce structural loading for wind farms application.

The desire to enhance our knowledge concerning the study of rotary wing aerodynamics has spurred researchers, scientists, and engineers to develop effective tools in both the experimental and numerical fields. These tools were essential to optimise the design process of novel machines or infrastructures characterised by configurations of single or multiple rotating lifting bodies, particularly aiming in improve their performance, structural dynamics, handling qualities, and acoustic impacts. Experimental activities in this research field were mainly based on wind tunnel tests performed over test rigs reproducing the dynamics of real rotors blades. Particular effort was devoted to the development of pitching airfoil test rigs capable of reproducing both the dynamics and the real flow conditions of a rotor blade section. Wind tunnel campaigns using these test rigs were highly useful for the study of the dynamic stall process, which represents a phenomenon that negatively influences both the aerodynamic performance characteristics and structural dynamics of helicopters and wind turbine rotors. A step forward concerning the investigation of classical aerodynamic phenomena characterising a rotor was achieved using whirl towers test rigs. These experimental rigs reproduced the complete mechanical system of real rotor hubs but on a smaller scale, thus showing a high manufacturing complexity related to the miniaturisation of their main constitutive parts, such as hinges, pitch rods, and actuators. Nevertheless, these test rigs enabled researchers to operate under monitored laboratory conditions, thus contributing to the achievement of more detailed insights with respect to real operative stands and to the highly unsteady flow features characterising rotating blades. A particular boost to the knowledge of flow physics in this research area was provided by recent developments in the field of optical measurement techniques. The use of Particle Image Velocimetry (PIV), particularly in stereoscopic or time-resolved modes, for instance, has enabled researchers to fully describe the fine details of the dynamic stall process over retreating rotor blades or to analyse several flow mechanisms typical of blade–vortex interactions (BVI). Moreover, recent advances in Pressure-Sensitive Paint (PSP) or Infrared Thermography (IR) measurements enabled researchers to accurately investigate transient aerodynamic phenomena, such as flow separation or the laminar to turbulent boundary layer transition, occurring over rotating blade surfaces without using intrusive probes.

Numerical tools were also effective in recent studies concerning rotary wing aerodynamics. Indeed, the advances achieved in recent years in the field of high-performance computing has allowed an increase in the use of high-fidelity Computational Fluid Dynamics (CFD) solvers for investigating the complex interactional aerodynamics phenomena typical of rotary wing machines. These solvers, based on a finite-volume implementation of the Reynolds Averaged Navier–Stokes (RANS) equations, enables researchers to manage moving block-structured grids, particularly using the Chimera technique, thus easily performing complex simulations of vehicles characterised by multiple rotors and lifting bodies. Consequently, these numerical tools were successfully employed for aerodynamics studies of helicopters and complex rotorcraft vehicles, such as tiltrotors or compounds. Moreover, high-fidelity CFD tools were also widely used for wind turbines and turbomachinery simulations. High-fidelity solvers were thoroughly validated in the field of rotary wing aerodynamics research, as their solutions present a quite accurate agreement with wind tunnel measurements. Nevertheless, despite the advances in high-performance computing, time-accurate RANS simulations still require very high computational effort in terms of time and resources in applications of rotary wing machines. For this reason, high-fidelity CFD tools are usually employed for a limited number of simulations of a well-defined configuration of such complex vehicles. As a matter of fact, the high-fidelity numerical approach to aerodynamics is still not suitable for the preliminary design process of novel rotary wing machines which require a huge number of simulations. Consequently, in recent years, the attention to mid-fidelity numerical approaches to rotary wing aerodynamics, combining numerical models with different accuracy, is growing among researchers and engineers working in this field. Mid-fidelity solvers typically represent a combination of a boundary value problem based on potential methods and a vortex particles model of vorticity for the flow. In particular, the vortex particle method (VPM) is a grid-free model suitable to accurately reproduce the strong aerodynamic interactions occurring among wakes that are typical of complex rotary wing machines, such as eVTOLs or wind farms. Thus, a mid-fidelity numerical approach to rotary wing aerodynamics simulations represents an optimal trade-off between accuracy and computational effort. Indeed, the capability of mid-fidelity tools to obtain solutions for complex rotary wing vehicles in high agreement with experimental results, but with a limited computational time with respect to high-fidelity CFD solvers, have opened a new scenario in the design process of novel rotary wing machines.

The goal of this Special Issue is to collect experimental and numerical studies showing recent advancements in the study of rotary wing aerodynamics. Due to the transversal content of this topic, the Special Issue attracted works from both aerospace engineering and wind energy specialists. In particular, the Special Issue contains six articles. Three articles deal with rotorcraft aerodynamics applications, i.e., two numerical studies on innovative rotorcraft configurations such as eVTOL aircraft [1] and tiltrotors [2] and an experimental-numerical study aiming to study the fine wake details of a helicopter's main rotor [3]. Two articles deal with wind turbine aerodynamics applications, i.e., an experimental activity aimed to develop both aerodynamic and structural strategies to design an experimental model of a wind turbine rotor [4] and a numerical study aiming at the duct optimisation of a vertical-axis wind turbine finalised to power enhancement [5]. The last article describes a trade-off activity between rotorcraft and wind energy research fields, showing the results of piloted simulations to determine possible risks associated to wind turbine interactions in rotorcraft operations [6]. The next section includes a brief overview of these articles, pointing out their main findings and their novelties with respect to the state-of-the-art in the field of rotary wing aerodynamics research.

2. Special Issue Articles' Short Review

The first article collected in the Special Issue by Piccinini et al. [1] describes a numerical activity aimed at performing a systematic study of the aerodynamic interactions between two propellers, with applications to eVTOL aircraft flight conditions. As previously mentioned, these aircraft represent the greatest novelty in the aeronautical field developed in recent years. These aircraft configurations are widely investigated throughout the world, and even though very different layouts are under development, a common key feature of their architecture is represented by multiple propellers positioned in side-by-side and tandem configurations over single or dual-lifting surfaces. Consequently, a systematic study of the basic flow mechanisms involved in the aerodynamic interaction between two propeller represents a milestone in the development and optimisation of these novel aircraft configurations. This work provides interesting guidelines for eVTOL design. Indeed, the mid-fidelity numerical solver employed in this study was suitable to capture the fine details of the interactional flow field characterising the investigated propeller configurations. In addition, the numerical investigation provided a quantitative indication about the interactional effects on the propellers' aerodynamic performance by highlighting the propellers' performance losses due to their mutual separation distance and the different degree of overlap between their rotor disks. Moreover, this work showed that numerical results with a high level of accuracy compared to experiments could be obtained by a solver requiring a very low computational effort with respect to high-fidelity CFD tools. Thus, the outcomes of this activity confirmed for scientific and industrial communities the suitability of a mid-fidelity numerical approach to aerodynamics for the preliminary design and optimisation of novel eVTOL aircraft configurations which require a huge number of simulations to investigate the several phases characterising the flight mission of these vehicles.

The second article collected in this Special Issue by Muggiasca et al. [4] is focused on the investigation of "best practices" to be adopted to perform experiments on scaled wind turbine blade models. As a matter of fact, experimental activities have a key role in the investigation and development of wind energy technologies. In particular, the present article is focused on the strategies of designing a scaled wind turbine blade model suitable for obtaining a fluid–structure interaction comparable to real machine blades. In particular, applications to rotor blade models for wind tunnel tests as well as natural laboratory tests were considered by the authors. This work considers both the aerodynamic and structural design of a floating wind turbine blade model, showing that non-Froude performance scaling can favour the reproduction of the full-scale rotor's aerodynamic behaviour and can improve its agreement with the real wind turbine's thrust coefficient while also preserving the power coefficient shape.

The third article collected in this Special Issue by De Gregorio et al. [3] describes an experimental and numerical activity aimed at investigating the wake of a helicopter rotor in the hovering condition in detail. This article represents a successful attempt to combine a quite modern experimental technique such as PIV with a classical numerical method, i.e., the free-wake Boundary Element Methodology (BEM), to study a complex problem in the field of rotary wing aerodynamics, i.e., the investigation of the vortex decay process during the downstream convection of a rotor wake in hovering conditions. In particular, the thorough comparison between experimental and numerical results highlights the degree of accuracy provided by a lower-order numerical method such as BEM to capture the trajectory of the filament vortex and in the interest of the complex interactions occurring between the tip vortices issued by rotating blades. Consequently, the outcomes of this article provide quite interesting indications to validate low-order numerical methods for the investigation of complex aerodynamics problems typical of classical rotary wing applications. Indeed, as previously stated, the validation of low-order numerical tools with increasing accuracy represents an essential task for the development of novel rotorcraft vehicle configurations.

The fourth article collected in this Special Issue by Ranjbar et al. [5] turns readers back to the investigation of wind turbine aerodynamics. In particular, the article is focused on the aerodynamic performance optimisation of a quite interesting machine in the wind energy field, i.e., a vertical-axis-ducted wind turbine. The article illustrates in detail a successful numerical activity aimed to optimise the duct's geometry used to collect the flow over a wind turbine rotor for the purposes of power enhancement. This work illustrates an effective example of the capabilities of a modern numerical tool to perform all the steps required for an aerodynamics geometry optimisation problem, from geometry and mesh generation to solver setup and simulations execution. In particular, the numerical results presented in this article clearly highlight the need for a high-fidelity aerodynamic solver based on Navier–Stokes equations equipped with a suitable turbulence model to properly describe the vortical structure's evolution typical of the deep dynamic stall phenomena occurring on wind turbine rotor blades.

The fifth article collected in this Special Issue by Savino et al. [2] turns the reader's attention back again to the rotorcraft research field, particularly with a focus on an interdisciplinary activity connecting aerodynamics to structural dynamics. The main goal of this work is to underline the importance of a more accurate aerodynamic numerical model for aeroelastic studies of complex rotary-wing aircraft. The article illustrates a novel numerical tool obtained by coupling a VPM-based, mid-fidelity aerodynamic solver to a multibody dynamics code. This tool is completely open source. The numerical results shown in this article, obtained by coupled simulations reproducing a full tiltrotor during a transient rolling manoeuvre, confirm that the use of VPM for the modelling of rotating blade wakes introduces an apparent benefit for the evaluation of rotor aerodynamic loads and consequently improves the aeroelastic assessment of rotary-wing aircraft configurations typically characterised by complex interactional aerodynamic features. Moreover, the quite limited computational effort shown by this coupled numerical tool supports the suitability of this enhanced approach to aerodynamics finalised to obtain higher accuracy in the preliminary design of novel rotary-wing vehicles.

The sixth article by Strbac et al. [6] brings readers to an interdisciplinary scenario where wind turbine aerodynamics plays a significant role in rotorcraft piloting. This work deals with helicopter operations within an offshore wind farm environment, with particular focus on the interaction between wind turbine wakes and helicopter flight paths. The approach used to investigate this problem is quite novel and interesting. Indeed, the use of high-fidelity CFD methods for modelling the far- and near-wake flow field of a wind turbine superimposed as input to a flight simulator campaign enabled the authors to obtain realistic information for piloting helicopters in the maritime environment of a wind farm. As a matter of fact, the outcomes of this work provide significant and appreciable indications for avoiding potential risks for helicopter operations in these complex scenarios by suggesting the proper size of the flight corridor and a sufficient lateral safety clearance near offshore wind turbines.

3. Conclusions

The articles published in this Special Issue cover a wide range of research topics in the main key areas of rotary wing aerodynamics. High-quality experimental and numerical techniques highlighted how to deal with such complex physical problems in a modern and effective way. Furthermore, these works provide insights that must be considered very useful to enhance knowledge in this discipline and to favour innovative developments both in rotorcraft and wind energy research areas. Indeed, the investigated topics are widely investigated by the scientific and industrial research community, and I thoroughly believe that the collected works will spur further authors to deepen the findings disseminated by the present Special Issue.

Generally, I would like to thank all the authors for the precious work devoted to the activities disseminated in the present Special issue and to the manuscript's preparation. The collected works meet the high-quality standard of *Energies*, and I hope that future Special Issues of this journal could consolidate the interest of the research community concerning rotary wing aerodynamics, particularly deepening topics related to electrical air mobility and wind farms applications.

Funding: This work has received no funding.

Institutional Review Board Statement: Not applicable.

Informed Consent Statement: Not applicable.

Data Availability Statement: Not applicable.

Acknowledgments: I would like to acknowledge Assistant Editor and the *Energies* Editorial Staff for their precious support during the development of this Special Issue.

Conflicts of Interest: The author declares no conflict of interest.

References

1. Piccinini, R.; Tugnoli, M.; Zanotti, A. Numerical Investigation of the Rotor-Rotor Aerodynamic Interaction for eVTOL Aircraft Configurations. *Energies* **2020**, *13*, 5995. [CrossRef]
2. Savino, A.; Cocco, A.; Zanotti, A.; Tugnoli, M.; Masarati, P.; Muscarello, V. Coupling Mid-Fidelity Aerodynamics and Multibody Dynamics for the Aeroelastic Analysis of Rotary-Wing Vehicles. *Energies* **2021**, *14*, 6979. [CrossRef]
3. De Gregorio, F.; Visingardi, A.; Iuso, G. An Experimental-Numerical Investigation of the Wake Structure of a Hovering Rotor by PIV Combined with a Γ_2 Vortex Detection Criterion. *Energies* **2021**, *14*, 2613. [CrossRef]
4. Mugglasca, S.; Taruffi, F.; Fontanella, A.; Di Carlo, S.; Belloli, M. Aerodynamic and Structural Strategies for the Rotor Design of a Wind Turbine Scaled Model. *Energies* **2021**, *14*, 2119. [CrossRef]
5. Ranjbar, M.H.; Rafiei, B.; Nasrazadani, S.A.; Gharali, K.; Soltani, M.; Al-Haq, A.; Nathwani, J. Power Enhancement of a Vertical Axis Wind Turbine Equipped with an Improved Duct. *Energies* **2021**, *14*, 5780. [CrossRef]
6. Štrbac, A.; Greiwe, D.H.; Hoffmann, F.; Cormier, M.; Lutz, T. Piloted Simulation of the Rotorcraft Wind Turbine Wake Interaction during Hover and Transit Flights. *Energies* **2022**, *15*, 1790. [CrossRef]

Article

Numerical Investigation of the Rotor-Rotor Aerodynamic Interaction for eVTOL Aircraft Configurations

Riccardo Piccinini, Matteo Tugnoli and Alex Zanotti

Dipartimento di Scienze e Tecnologie Aerospaziali, Politecnico di Milano, via La Masa 34, 20156 Milan, Italy; riccardo.piccinini@mail.polimi.it (R.P.); matteo.tugnoli@polimi.it (M.T.)
* Correspondence: alex.zanotti@polimi.it

Received: 21 October 2020; Accepted: 12 November 2020; Published: 17 November 2020

Abstract: The rotor-rotor aerodynamic interaction is one of the key phenomena that characterise the flow and the performance of most of the new urban air mobility vehicles (eVTOLs) developed in the recent years. The present article describes a numerical activity that aimed to the systematic study of the rotor-rotor aerodynamic interaction with application to the flight conditions typical of eVTOL aircraft. The activity considers the use of a novel mid-fidelity aerodynamic solver based on vortex particle method. In particular, numerical simulations were performed when considering two propellers both in side-by-side and tandem configuration with different separation distances. The results of numerical simulations showed a slight reduction of the propellers performance in side-by-side configuration, while a remarkable loss of thrust in the order of 40% and a reduction of about 20% of the propulsive efficiency were found in tandem configuration, particularly when the propeller disks are completely overlapped. Moreover, the flow field analysis enabled providing a detailed insight regarding the flow physics involved in such aerodynamic interactions.

Keywords: rotary-wing aerodynamics; rotor interaction; eVTOL aircraft; computational fluid dynamics; vortex particle method

1. Introduction

In recent years, a great interest and development effort has been devoted towards the design of unconventional VTOL aircraft based on electric distributed propulsion (eVTOLs) with the aim to create a novel concept of urban air mobility to be considered as an effective alternative to ground transportation in overcrowded metropolitan areas [1]. The development of these new aircraft architectures combining in a single vehicle aerodynamic elements typical of different classical configurations, such as fixed lifting surfaces, lifting rotors, and thrusting propellers, pose unprecedented challenges to engineers in several areas. In particular, even if the aircraft architectures that are designed by the companies are rather diverse, the rotor-rotor interaction represents, from an aerodynamic standpoint, one of the novel key phenomena that characterise the flow around most eVTOLs as well as their performance, handling qualities, and noise. Indeed, the common feature that characterises eVTOLs design is the use of multiple propellers, as illustrated by the layout of few examples of eVTOLs aircraft developed in the last years shown in Figure 1. The multiple propellers mounted on wings are typically close to each other, as can be observed from the layout of the novel aircraft designed by Archer that is shown in Figure 1a and by the Vahana aircraft architecture designed by A³ by Airbus LLC [2] shown in Figure 1b. Moreover, the propellers are often arranged on two lifting surfaces with different longitudinal separation distance and typically present a certain region of overlapping between the rotor disks, as can be observed from the layout of the Bell-Nexus 6HX designed by Bell shown in Figure 1c and by the S4 aircraft that were designed by Joby Aviation shown in Figure 1d. Therefore, two main types of rotor-rotor aerodynamic interaction can be outlined as the more interesting for eVTOL applications, i.e., with the propellers in side-by-side and tandem configurations.

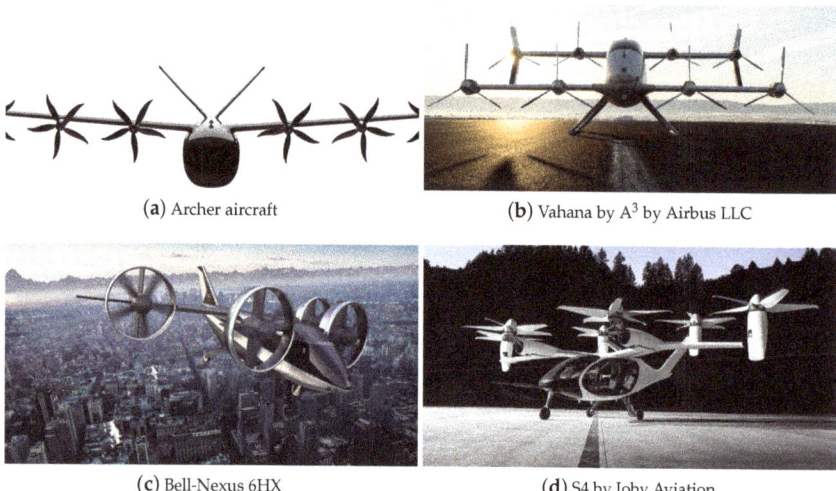

(a) Archer aircraft   (b) Vahana by A³ by Airbus LLC

(c) Bell-Nexus 6HX   (d) S4 by Joby Aviation

Figure 1. Examples of eVTOLs aircraft architectures (from https://evtol.news/aircraft): (**a**) Archer aircraft, https://www.flyarcher.com/; (**b**) Vahana by A³ by Airbus LLC, https://acubed.airbus.com/projects/vahana/; (**c**) Bell-Nexus 6HX by Bell, https://www.bellflight.com/products/bell-nexus; (**d**) S4 by Joby Aviation, https://www.jobyaviation.com/.

In the recent years, the investigation of these kinds of rotor-rotor aerodynamic interaction has begun to gather interest in the scientific community both in experimental and numerical field, in particular for applications of multirotor drones. For instance, the experimental work by Zhou et al. [3] investigated the interactional effect of the distance between two small UAV propellers in side-by-side configuration for hover conditions. This work shows that a negligible reduction of the interacting propellers performance is obtained in such test conditions, while a high level of unsteady load fluctuations is achieved, decreasing the lateral distance between the rotor disks. A similar test configuration, when considering a side-by-side configuration of two mini-drone rotors in hover, was investigated in the experimental work by Shukla and Komerath [4]. This work shows, by means of stereo Particle Image Velocimetry (PIV) measurements, an increase of the rotor wake interactions decreasing the rotor spacing and Reynolds number. Concerning the investigation of the aerodynamic interaction of rotors in tandem the majority of the works dealt with coaxial rotors configuration. For instance, wind tunnel tests were performed by Shukla et al. [5] in order to study the aerodynamic interaction of two coaxial rotors considering particularly the effects of Reynolds number and of advance ratio. In particular, this work showed that, for low Reynolds, the upper rotor provides a higher figure of merit due to swirl recovery. Moreover, Brazinskas et al. [6] investigated, by means of loads and torque measurements, the performance of two co-axial rotors when also considering partially overlapped conditions between the rotors disks, but with longitudinal distance below a rotor radius.

Despite this effort, there is a certain lack in literature of a systematic study aimed in order to obtain parametric data on the different types of rotor-rotor interactions occurring in flight conditions typical of the eVTOL vehicles. In particular, there is a limited effort in the scientific literature regarding the investigation of aerodynamic interaction between rotors in tandem characterised by large longitudinal distance and a different degree of overlapping region between rotors that are typical characteristics of several eVTOL architectures. Moreover, there is a lack concerning research works that aimed to analyse the aerodynamic interaction of rotors in both tandem and side-by-side configurations for cruise flight conditions that are typical of eVTOLs aircraft.

Computational fluid dynamics (CFD) simulations represent a valuable tool for the investigation of this complex aerodynamic interactions. Indeed, high-fidelity CFD simulations were used in the work by Yoon et al. [7] to investigate the the performance and efficiency of multi-rotor systems with particular effort on the study of the effects of the separation distance between rotors, fuselage, and wings. Nevertheless, time-accurate URANS simulations of multi-rotor configurations still require a huge computational effort. Consequently, high-fidelity CFD simulations can not be considered to be the suitable tool for a systematic study of the different kinds of rotor-rotor interactions involved in eVTOLs aerodynamics. On the other hand, a mid-fidelity numerical approach combining different models and characterised by a low computational effort represents the best option for providing parametric data on the different types of rotor-rotor interactions and exploring a comprehensive range of space parameters. Several mid-fidelity aerodynamic solvers were developed in recent years with successful application to the study of rotary wing aerodynamics. To cite an example, DLR developed UPM code [8], an unsteady panel, and free-wake code originally intended for aeroacoustic simulations of helicopters but recently applied on arbitrary complex configurations as compound rotorcraft [9]. In particular, the use of vortex particle method (VPM) [10,11] for wake modelling enabled obtaining a better representation of the rotor aerodynamic characteristics and to capture the aerodynamic interactions between several bodies. Indeed, recent literature shows several works employing the VPM for the simulations of rotorcraft applications [12–15]. Concerning rotor-rotor interactions, Alvarez and Ning [16] used a mid-fidelity aerodynamic code based on VPM in order to simulate the side-by-side configuration of two propellers in hover investigated by Zhou et al. [3], finding good agreement with experiments for both the rotor performance and for the representation of the interacting flow fields.

A novel medium fidelity aerodynamic open-source software, called DUST (https://www.dust-project.org/), has been recently developed by Politecnico di Milano as the result of a collaboration with A^3 by Airbus LLC. The code was thoroughly validated against experiments and high fidelity CFD on different rotorcraft configurations from simpler rotor-wing test cases to a full eVTOL vehicle. In particular, a recent work [17] showed that DUST simulations provide a quite good representation of both the performance and flow physics of a half-span tiltwing vehicle. These results were found when comparing the DUST simulations results with both wind tunnel data and high-fidelity CFD results. DUST was also used in a recent work [18] in order to simulate the aerodynamics of the full Vahana vehicle developed by A^3 by Airbus LLC characterised by two rows of four rotors in tandem configuration. A quite good agreement between DUST simulations results and both flight test data and high-fidelity CFD results was found in this work for the full Vahana vehicle flight performance, particularly for cruise conditions. Thus, the results obtained confirm the suitability of mid-fidelity simulations performed with DUST for the study of the complex aerodynamic interactions that characterise multi-rotors aircraft configuration as eVTOLs. Consequently, DUST can be considered in a mature state to be used for the investigation of the rotor-rotor aerodynamic interactions that is the object of the present work.

Indeed, the present work aimed to provide a comprehensive analysis of the rotor-rotor aerodynamic interaction by means of mid-fidelity numerical simulations performed with DUST. In particular, the numerical simulations investigated the aerodynamic interactions between two propellers both in side-by-side and in tandem configuration. The low computational effort that is required by the mid-fidelity solver used in this activity enabled to study several tandem configurations characterised by both low and large longitudinal distances between the propellers and several degrees of overlapping between the rotor disks. In particular, the present numerical activity was focused on the cruise flight condition of an eVTOL aircraft.

The paper is organized, as follows. A brief outline of the numerical approach implemented in DUST is presented in Section 2 with a particular insight on the description of the lifting line elements used for blade modelling and of the VPM implemented for wake modelling. Section 3 provides the validation of the parameters used for DUST simulations by means of comparison of the numerical results with experimental data available in literature for a propeller model test case in forward flight conditions. Section 4 describes the numerical model that was implemented for the study of the interacting propellers and the test cases analysed in the numerical activity. Section 5 presents the discussion of the main results obtained by mid-fidelity simulations in terms of propeller performance and flow physics involved in the problem. Conclusions are drawn in Section 6.

2. Numerical Approach Implemented in DUST

DUST was developed in order to obtain a fast, flexible, and accurate numerical solver that is suitable to perform aerodynamic simulations of complex aircraft configurations capable to provide a reliable representation of interactional aerodynamic phenomena. DUST is a flexible aerodynamics tool allowing for simulating models with different levels of accuracy. In particular, the solver integrates thick surface panels, thin vortex lattices and lifting lines to model solid bodies, while panels and vortex particles can be used to model the wake. The vortex particles wake is accelerated while using a Fast Multipole Method (FMM) [19] that was developed to obtain a robust and accurate model to simulate interactional aerodynamics phenomena, particularly for multiple wake interactions or for interactions of wakes with solid bodies. The implementation of DUST employed the use of the Object Oriented paradigms of the latest Fortran standards in order to obtain a high level of flexibility to model a complex aircraft configuration made of several components and to describe their motion. The aerodynamic solver is based on the Helmholtz decomposition of the velocity field, $\vec{u} = \vec{u}_\varphi + \vec{u}_\psi$, where $\vec{u}_\varphi$ and $\vec{u}_\psi$ are, respectively, the irrotational and solenoidal contributions. A time-stepping algorithm alternating the solution of a three-dimensional boundary element method for $\vec{u}_\varphi$ and the Lagrangian time evolution of the rotational contribution of the velocity $\vec{u}_\psi$ is implemented for the solution advancing in time. In order to run the simulation the surface mesh only of the investigated object is required. Depending on the level of fidelity required, different aerodynamic elements can be used to discretise the model, in particular, lifting line elements, zero-thickness lifting surfaces, and surface panels. Surface panels are implemented while using a piecewise-uniform distribution of doublets and sources, according to a formulation for the velocity potential defined by Morino [20]. Moreover, zero-thickness surfaces of vortex lattice elements can be used in order to model thin lifting bodies. The result of the mixed potential-velocity formulation for the boundary element problem is a linear system where the unknowns are the doublet distribution intensity on the surface panels and the equivalent doublet intensity of the vortex lattice elements. In the following, a more detailed description of the lifting line elements used in the present work for the propeller blades modelling and of the implemented VPM used for rotor wake modelling is provided. A more detailed description of the numerical approach that was implemented in DUST is provided in [17,18].

2.1. Lifting Line Elements

The lifting bodies with high aspect ratio, as, for instance, rotor blades are properly modelled by one-dimensional lifting line elements. These elements naturally include viscous effects modelling, since they rely on tabulated aerodynamic performance of the two-dimensional sections of the modelled body. In particular, the aerodynamic tables are constituted by lift, drag and pitching moment coefficients as functions of the relative velocity direction and magnitude. Each lifting line element is constituted by a vortex ring along with its trailing vortices and the last line vortex are released in the wake aligned to the spanwise direction. The intensity Γ of the vortex ring, and therefore of the lifting line, is calculated through a fixed point algorithm that solves a nonlinear problem, connecting the lifting line elements intensity to the tabulated aerodynamic coefficients of the lifting sections. With this aim, DUST formulation implements both a loosely-coupled Γ-method [21] and a α-method [22] solver. The first method is based on the equivalence of the semi-empirical equation for the sectional lift with its analytical expression from the Kutta-Joukowski theorem,

$$\frac{1}{2}\rho |U_{rel,i}|^2 c_i c_{\ell,i}(\alpha_i(\Gamma_k)) = -\rho |U_{rel,i}|\Gamma_i, \qquad (1)$$

where c_i is the ith section chord, $c_{\ell,i}(\alpha_i)$ its lift curve, with α_i and $U_{rel,i}$ being the incidence angle and the relative velocity calculated at the control point of the ith lifting line. When considering that the lifting line element is positioned at 1/4 of the chord, the control point is evaluated at 3/4 of the chord [21,22]. In the latter method, the incidence angle resulting from the velocity field that is induced by all of the wake elements, including the particles and all the different lifting line elements constituting the model is considered as the input to find the corresponding aerodynamic coefficients in the two-dimensional tabulated aerodynamic data,

$$\alpha_i = \operatorname{atan2}(U_{rel,i}(\alpha_k) \cdot \hat{n}_i, U_{rel,i}(\alpha_k) \cdot \hat{t}_i). \qquad (2)$$

The nonlinear problem of computing the loads on lifting lines is solved through an iterative procedure that considers their mutual interference by means of the use of the Kutta–Joukowski theorem in order to evaluate the circulations of the elements from their lift.

2.2. Vortex Particle Method

The wake shed from the trailing edges by lifting bodies can be represented by vortex particles. The vortex particle [10,11] numerical modelling of the wake was introduced in order to provide a more robust representation of the interactional aerodynamics of both rotorcraft and complex aircraft configurations. The vortex particles method (VPM) is a Lagrangian grid-free method describing the wake evolution through the rotational component of the velocity field $\vec{u}_\psi$ by means of material vortex particles used to obtain the approximate vorticity field, as follows,

$$\vec{\omega}^h(\vec{r},t) = \sum_{p=1}^{N_p} \vec{\alpha}_p(t) \zeta\left(\vec{r}-\vec{r}_p(t); R_p\right), \qquad (3)$$

where $\vec{r}_p(t)$ is the position, $\vec{\alpha}_p(t)$ the intensity, and R_p the radius of the p-th vortex particle, while $\zeta(r)$ is the cut-off function considering the vorticity distribution induced by each particle. By substituting (3) in the equation of the dynamics of vorticity,

$$\frac{D\vec{\omega}}{Dt} = \vec{\omega} \cdot \nabla \vec{u} + \nu \nabla^2 \vec{\omega}, \qquad (4)$$

the dynamical equations for the intensity $\vec{\alpha}_p(t)$ and position $\vec{r}_p(t)$ of all the material vortex particles to be integrated in time can be obtained, as follows,

$$\begin{cases} \dfrac{d\vec{\alpha}_p}{dt} = \vec{\alpha}_p \cdot \nabla \vec{u}(\vec{r}_p(t),t) + \nu\,"\nabla^2 \vec{\alpha}_p" \\[4pt] \dfrac{d\vec{r}_p}{dt} = \vec{u}(\vec{r}_p(t),t). \end{cases} \qquad (5)$$

The viscosity diffusion term "$\nabla^2 \vec{\alpha}_p$" is calculated using the particle strength exchange method (PSE) that approximates the Laplacian operator acting on the vorticity field with an integral operator, as was described in [11].

The mathematical formulation used in the solver relies on the Helmholtz decomposition of the velocity field $\vec{u} = \vec{u}_\varphi + \vec{u}_\psi$. The irrotational velocity $\vec{u}_\varphi$ is induced by the free stream velocity, by the singularity distributions of the source and doublet on the body surface and by the wake panels, while the rotational velocity $\vec{u}_\psi$ is induced by the vortex particles. Moreover, the solenoidal constraint on the rotational velocity, $\nabla \times \vec{u}_\psi = \vec{0}$, is used in order to define the vector potential $\vec{\psi}$, s.t. $\vec{u}_\psi = \vec{\nabla} \times \vec{\psi}$. Consequently, the Poisson's equation is obtained for $\vec{\psi}$,

$$-\nabla^2 \vec{\psi} = \vec{\omega}, \qquad (6)$$

considering the gauge condition $\nabla \cdot \vec{\psi} = 0$, the vorticity field definition $\vec{\omega} = \nabla \times \vec{u}$ and the vector identity $\nabla \times \vec{u}_\varphi = \nabla \times \nabla \varphi = 0$. The Poisson's equation solution (6) reads

$$\vec{\psi}(\vec{r},t) = \int_{V_0} G(\vec{r},\vec{r}_0)\, \vec{\omega}(\vec{r}_0,t) dV_0, \qquad (7)$$

$$\vec{u}_\psi(\vec{r},t) = \int_{V_0} \vec{K}(\vec{r},\vec{r}_0) \times \vec{\omega}(\vec{r}_0,t) dV_0 \qquad (8)$$

where $G(\vec{r},\vec{r}_0)$ is the Green's function of the Laplace equation and $\vec{K}(\vec{r},\vec{r}_0)$ represents its gradient with respect to the first argument.

Substituting the definition of the discretized vorticity field of the particles (3) into Equation (8), the contribution of velocity induced by the particles can be obtained, as follows,

$$\vec{u}_\psi^h(\vec{r},t) = \sum_{p=1}^{N_p} \vec{K}^h(\vec{r} - \vec{r}_p(t)) \times \vec{\alpha}_p(t). \qquad (9)$$

The discrete kernel $\vec{K}^h(\vec{r} - \vec{r}_p(t))$ have to be consistent with the selected cutoff function ζ. The cutoff function in the singular vortex particle method is a Dirac delta function and the Biot–Savart kernel is retrieved. In DUST implementation the selected cutoff function ζ leads to the Rosenhead-Moore kernel,

$$\vec{K}^h(\vec{x},\vec{y}) = -\frac{1}{4\pi} \frac{\vec{x} - \vec{y}}{(|\vec{x} - \vec{y}|^2 + R_v^2)^{3/2}}, \qquad (10)$$

a regular kernel fitting naturally in the Cartesian fast multipole method (FMM) [19,23]. The induced rotational velocity $\vec{u}_\psi$ have to be accounted in the material objects convection and in the right-hand side of the linear system of equations for the potential velocity. Moreover, the velocity field gradient is calculated in order to evaluate the vortex stretching-tilting term with the FMM. Indeed, this term is a function of both the vortex intensities and particles distance in particle-to-particle interactions [11].

3. Validation of the DUST Simulations Parameters for a Propeller Test Case in Forward Flight Conditions

A validation study was performed by comparison with experimental data regarding a propeller model in forward flight in order to validate the DUST simulations parameters to be used for the simulation of the interacting propellers. The considered experimental data were obtained by McCrink and Gregory [24] for the APC thin-electric 10 × 7 propeller in forward-flight conditions. A numerical model of the APC 10 × 7 propeller was built for DUST simulations considering the airfoil geometry and the chord and twist distributions that were provided in the work by McCrink and Gregory [24]. Each of the two blades of the propeller was modelled using lifting lines elements, naturally including the viscosity contributions to aerodynamic loads through tabulated sectional aerodynamic data. The tabulated data of the blade airfoils were computed by XFOIL simulations [25], before stall angle of attack, while the Viterna method [26] was used in order to obtain the post-stall behaviour of the two-dimensional aerodynamic loads coefficients curves to cover the range ±180° of angle of attack. DUST simulations reproduce a sweep of advance ratio J, defined as $J = V_\infty/(nD)$, where V_∞ is the free-stream velocity, $n = RPM/60$ is the propeller rotational velocity, and D is the propeller diameter of 0.254 m. The rotational speed of the propeller was fixed to 9200 RPM corresponding to a tip Mach number (M_t) of 0.36 and a Reynolds number $Re_D = 1.5 \cdot 10^6$, where $Re_D = V_{70\%} D/\nu$ and $V_{70\%}$ is the effective velocity at 70% blade span. The propeller thrust (T) coefficient C_T, the torque (Q) coefficient

C_Q, the power (P) coefficient C_P, and the propulsive efficiency η considered throughout the paper for the results discussion are defined as

$$C_T = \frac{T}{\rho n^2 D^4}, \qquad C_Q = \frac{Q}{\rho n^2 D^5}, \qquad C_P = \frac{P}{\rho n^3 D^5}, \qquad \eta = J\frac{C_T}{C_P}. \qquad (11)$$

A dependence study was performed for this test case due to the availability of experimental data in order to evaluate the optimal spatial and temporal discretization parameters for the propeller simulations. The full description of this study is reported in [27]. With this aim, numerical simulations were performed for $J = 0.6$ fixing the spatial discretization to 20 lifting line elements for each blade and changing the time step throughout a simulation length of 10 rotor revolutions (N_{rev}). A minimum error of the computed C_T with respect to the experimental value was found for a time discretization that corresponds to $5°$ of blade azimuthal angle for each rotor revolution. Halving the time-step to $2.5°$ the variation of the C_T error was negligible. Subsequently, simulations were repeated fixing the time-step to $5°$ of blade azimuthal angle for each rotor revolution and increasing the number of lifting line elements to model the blades. A minimum error of the computed C_T with respect to the experimental value was found while using 40 lifting line elements. Consequently, numerical simulations for a sweep of J were performed while using the optimal parameters found from the spatial and temporal dependence study. In particular, simulations were advanced in time with a discretization of $5°$ of blade azimuthal angle for each rotor revolution, while each blade was modelled using 40 lifting line elements.

Figure 2a shows the time histories of the thrust coefficient C_T calculated by DUST throughout a simulation length of 10 rotor revolutions (N_{rev}) for the APC 10×7 propeller at different advance ratios J. The curves behaviour shows that after five rotor revolutions the computed thrust coefficients reach a steady value for all of the advance ratio J, thus confirming that the number of rotor revolutions used for the simulations is quite enough to reproduce a fully developed wake of the propeller and obtain converged values of the propeller performance coefficients.

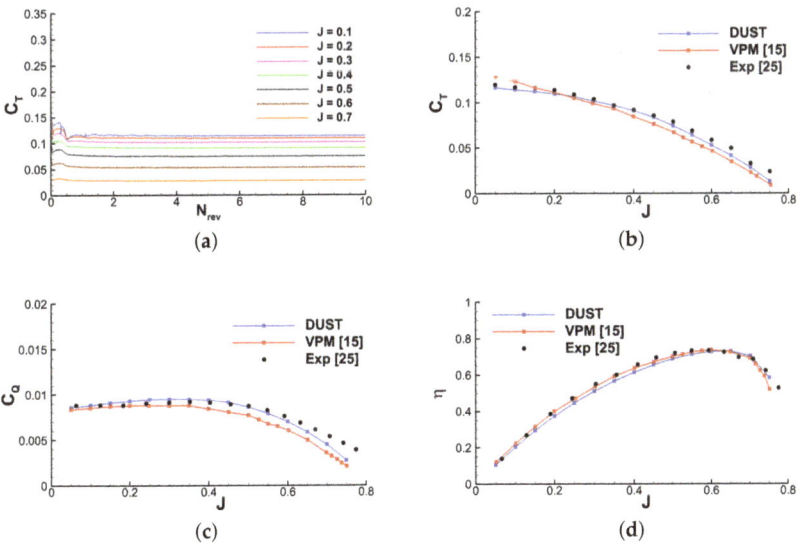

Figure 2. Comparison of the results obtained for the APC 10×7 propeller, $Re_D = 1.5 \times 10^6$, $M_t = 0.36$. (**a**) Time histories of the thrust coefficient C_T calculated by DUST as a function of the number of rotor revolutions N_{rev}; (**b–d**) comparison of the DUST simulations results with the experimental data from McCrink and Gregory [24] and the numerical simulations results from Alvarez and Ning [16].

Figures 2b–d show the comparison of DUST simulations results with the experimental data from McCrink and Gregory [24] and the numerical results from Alvarez and Ning [16] obtained on the same test case with a VPM-based code. The DUST simulations results were obtained by averaging the C_T and C_P that were calculated over the last two rotor revolution.

The behaviour of the C_T, C_Q, and propulsive efficiency η curves computed by DUST simulations is in quite good agreement with the experimental data. In particular, the DUST simulations showed a higher accuracy with respect to the results obtained by a similar VPM-based solver [16] for the evaluation of C_T in the range of advance ratio between $J = 0.4$ and $J = 0.65$ and at low advance ratios (see Figure 2b). Concerning C_Q evaluation, DUST showed the same quite good matching with experimental data for almost the whole range of advance ratios tested, with some discrepancies only observed at high advance ratios (see Figure 2c). The quite good agreement with experimental and numerical data available in literature observed in Figure 2d for the propulsive efficiency confirms that the parameters used to build the numerical model and run the simulations in DUST can be considered to be suitable for the study of a propeller model performance in forward-flight conditions. Therefore, the same parameters that were used for the APC propeller simulations in terms of time and space discretisations were used for the numerical study of the interacting propellers.

4. Numerical Model of the Interacting Propellers

The model used for the study of the rotor-rotor aerodynamic interactions is a three-bladed propeller equipped with a Varioprop 12C blade with a rotor radius R of 0.15 m. This hobby-grade model propeller was selected for this study, because is commercially available and provides dimensions that are suitable for performing experiments in a medium size wind tunnel for a thorough validation of the numerical investigation. The blade geometry was digitally created by 3D scanning of the blade model. CAD software was used in order to manage the generation of the blade geometry from the surfaces provided by the scanning system. Figure 3a shows the geometry of the blade where the coloured bar indicates an error below 0.1 mm between the reconstructed CAD geometry and the surfaces that were provided by the 3D scanner.

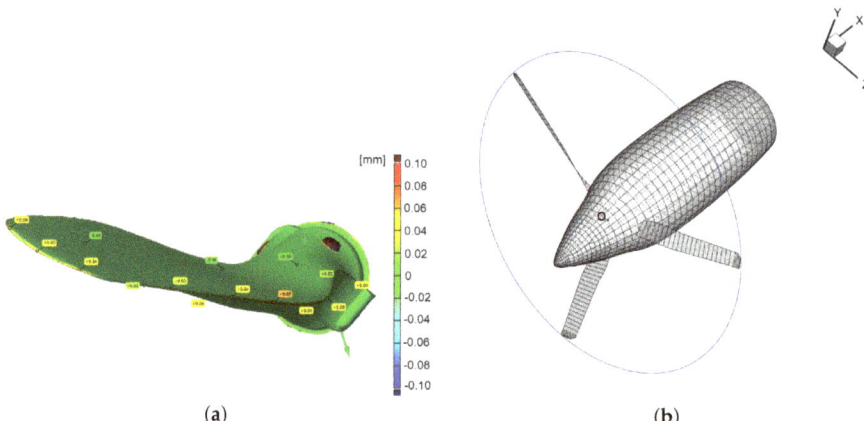

Figure 3. (a) Geometry of the Varioprop 12C propeller blade; the coloured bar indicates the error in mm between the reconstructed CAD geometry and the surfaces provided by the three-dimensional (3D) scanner. (b) Layout of propeller numerical model and reference system.

Figure 4 shows the twist, dihedral angle, and chord distributions along the blade radial coordinate (r). The propeller model used for this study is completed by a nacelle equipped with a 65 mm diameter spinner in order to reproduce the geometry of a eVTOL aircraft propeller. The airfoil sections and the nacelle geometry will be provided by request to the authors.

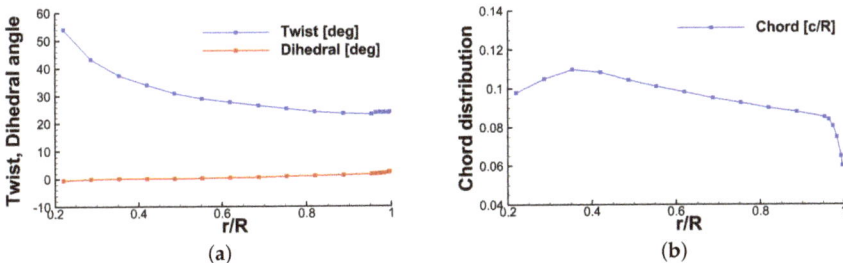

Figure 4. Geometrical description of the Varioprop 12C propeller blade. **a** Twist and dihedral angle distributions along the spanwise radial coordinate; **b** chord distribution along the spanwise radial coordinate.

For DUST simulations, the propeller blades were modelled while using 40 lifting lines elements each and the tabulated aerodynamic coefficients before stall were provided by XFOIL simulations [25] that were computed on the airfoil sections provided by the three-dimensional (3D) scanning along the span. The Viterna method [26] was used to obtain the post-stall behaviour of the sectional aerodynamic loads coefficients to cover the range between ±180° of angle of attack. The spinner-nacelle surface was modeled with 1212 surface panel elements. The layout of the numerical model, including the reference system used throughout the results discussion, is shown in Figure 3b. The origin (O) of the reference system $x - y - z$ is positioned on the center of the propeller disk, while the x axis is directed downstream and it is aligned with the free-stream velocity vector. In all of the simulations, the longitudinal axis of the nacelle is aligned with free-stream velocity vector.

Description of the Analysed Interacting Propellers Configurations

The numerical activity was focused on the study of two propellers in both tandem and side-by-side configurations in forward flight, with particular attention on the typical cruise flight velocity of eVTOL urban air mobility vehicles that can be considered in the order of 100 km per hour (i.e., V_∞ = 28 m/s). In all of the simulations, the rotational speed of both the propellers was fixed to 7000 RPM to reproduce the full-scale tip Mach number ($M_t = 0.32$) of a eVTOL aircraft in cruise [2]. The Reynolds number calculated on the base of the propeller disk diameter and on the rotational velocity evaluated at 70% R is $Re_D = 1.93 \times 10^6$. The blade pitch angle at 75% of the rotor radius was fixed to $\theta = 25.5°$ for both the interacting propellers. In order to reproduce cruise flight conditions of eVTOLs, the simulations were performed with the propellers aligned to the freestream velocity vector. In the following, the longitudinal distance along x axis between the rotor disks planes is defined as L_x, while the lateral distance between the propellers shaft axis is defined as L_y, as shown in the layout of the interacting configurations presented in Figure 5.

A preliminary simulation of the single propeller was performed in order to obtain the reference performance for comparison with the interacting configurations results.

The investigation of the side-by-side propellers interaction was focused on the advance ratio $J = 0.8$ that corresponds to a freestream velocity of V_∞ = 28 m/s, considered the target cruise velocity for eVTOLs. In particular, numerical simulations were performed for two counter-rotating propellers with rotor disks lying on the same $y - z$ plane ($L_x = 0$) at different lateral separation distances that range from $L_y = 2.05R$ (blade tips distance equal to 0.05R) to $L_y = 4R$.

The tandem interaction was investigated when considering two co-rotating propellers positioned with two different longitudinal distances between the rotor disks. In particular, simulations were performed with a longitudinal distance of 6 rotor radii ($L_x = 6R$) between the tandem propellers disks in order to reproduce the configuration of a Vahana-like aircraft (see Figure 1b). Moreover, in order to evaluate the effect of the longitudinal distance on the rotor-rotor interaction, simulations were performed also with $L_x = 2.5R$ reproducing the configuration of a more compact propulsive system that characterise for instance the Bell-Nexus 6HX aircraft architecture (see Figure 1c). A sweep along y axis was considered to evaluate the effect of the interaction due to a different overlapping area between the tandem propellers disks. In particular, the simulations reproduce a lateral sweep that ranges from the configuration where the propellers disks are completely ($L_y = 0$) to a separation distance between the propellers shaft axis of two rotor radii ($L_y = 2R$). Becasue of the low computational effort of the mid-fidelity approach, all of the tandem simulations were performed for a sweep of advance ratios J between 0.4 and 0.9 with a step of 0.1.

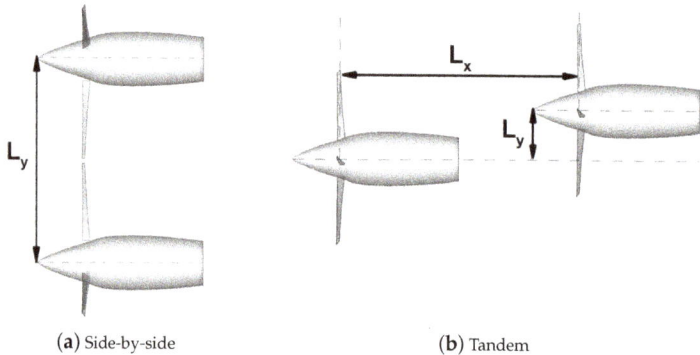

(a) Side-by-side (b) Tandem

Figure 5. Layout of the interacting propellers configurations investigated by numerical simulations.

Table 1 summarises the configuration parameters of the DUST simulations performed in this numerical activity. In particular, as done for the APC propeller numerical analysis, the interacting propellers simulations were performed for a length of 10 rotors revolutions with a time discretisation of 5° of blade azimuthal angle. A fully developed wake for the interacting propellers test cases consisted of around one million vortex particles. The computational time of the simulation of a single interacting configuration while using a workstation with a 18 cores processor was approximately 40 min for both the tandem and side-by-side test conditions.

Table 1. Summary of the configurations analysed by DUST simulations.

	RPM	$\theta_{75\%R}$	J	L_x [R]	L_y [R]
Single Prop	7000	25.5°	[0.4, 0.5, 0.6, 0.7, 0.8, 0.9]	-	-
Side-by-side Props	7000	25.5°	0.8	0	[2.05, 2.15, 2.25, 2.4, 2.5, 2.7, 3, 3.3, 3.5, 4]
Tandem Props	7000	25.5°	[0.4, 0.5, 0.6, 0.7, 0.8, 0.9]	[2.5, 6]	[0, 0.25, 0.5, 1, 1.5, 2]

5. Results and Discussion

5.1. Side-by-Side Propellers Configuration

In this section, the results that were obtained for the side-by-side propellers simulations are presented. In particular, the effect of the lateral distance between the rotor disks is discussed by means of comparison of the performance coefficients and by means of the analysis of the interacting flow field.

Figure 6a shows the time histories of the thrust coefficient C_T calculated for the counter-clockwise rotating propeller in side-by-side configuration (upper propeller of Figure 5a) at some different lateral distances L_y as compared to the results that were obtained from the single propeller simulation.

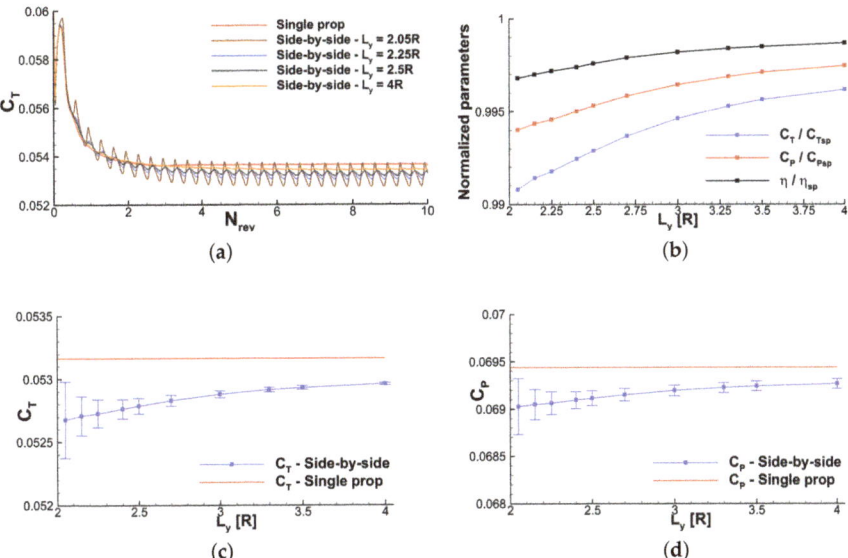

Figure 6. The results of the numerical simulations for the side-by-side interacting case, $\theta = 25.5°$, $M_t = 0.32$, $J = 0.8$. (**a**) Time histories of the thrust coefficient C_T computed for the single propeller and for the counter-clockwise rotating propeller (upper propeller of Figure 5a) in side-by-side configuration at different lateral distances L_y; (**b**) normalised thrust coefficient C_T, power coefficient C_P and propulsive efficiency η with respect to the single propeller parameters as function of the lateral distance L_y; and, (**c**,**d**) averaged thrust coefficient C_T and power coefficient C_P as function of the lateral distance L_y, errorbar corresponding to the standard deviation of the coefficients calculated over the last three rotor revolutions.

The C_T curves behaviour shows that after a transient of four revolutions, the results that were obtained for the side-by-side propellers simulations reach a periodic behaviour, while the single propeller simulation reaches a converged steady state value. The periodic behaviour obtained for the side-by-side simulations reflects the physics of the aerodynamic interaction between the propellers that is the object of this study, thus confirming the suitability of the numerical model considered to investigate this problem. The periodicity of the C_T curves is particularly evident in the last three revolutions of the simulations, thus the load coefficients considered in the following discussion were obtained as the averaged values that were calculated over this time interval.

Figure 6b shows the averaged thrust coefficient C_T, power coefficient C_P and propulsive efficiency η computed for the propeller in side-by-side configuration normalized with respect to the corresponding parameters evaluated from the single rotor simulation. The average performance of the propellers in side-by-side configuration are negligibly affected by the aerodynamic interaction, as can be observed in Figure 6b. Indeed, a loss of performance lower than 1% of the single propeller is observed for both thrust, power and propulsive efficiency when the propellers hubs are at a lateral distance $L_y = 2.05R$. Increasing the lateral distance L_y, the propeller resumes the performance of the single propeller configuration. The amount of thrust loss that was calculated in the present activity at high advance ratio is slightly lower than the outcomes of the works by Zhou et al. [3] and by Alvarez et al. [16] for side-by-side propellers at similar separation distance but in hover condition. Even if the effects on the average performance of the propellers can be considered negligible, an interesting effect of the aerodynamic interaction between the propellers is the fluctuation of the loads occurring during a rotor revolution. An indication of the loads fluctuations amplitude is provided by the standard deviation of the C_T and C_P computed over the last three revolutions plotted as errorbars in Figure 6c,d. This representation shows that the amplitude of the loads fluctuations is quite high when the lateral separation distance between the propellers is small. Indeed, when the distance between the propellers blades tips are equal to 0.05, a robust interaction between the tip vortices is expected, as will be discussed in the following analysis of the instantaneous flow fields. As the separation distance increases, the thrust and power coefficients for the side-by-side propellers approach the values of the single propeller configuration, while the load fluctuations amplitude decreases.

In order to provide a more detailed analysis of the local performance of the propeller blades at $L_y = 2.05R$, Figure 7 shows the difference of the sectional lift coefficient C_l and the effective angle of attack α_{eff} experienced by a propeller blade in side-by-side configuration with respect to the single rotor configuration that was computed during the last revolution.

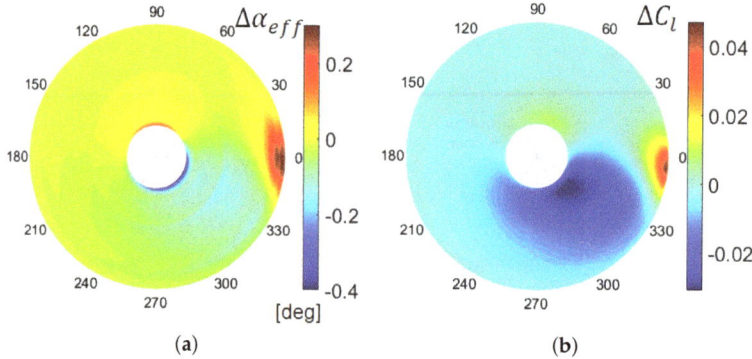

Figure 7. Variations of the effective angle of attack $\Delta\alpha_{eff} = \alpha_{eff} - \alpha_{eff_{sp}}$ and of the sectional lift coefficient $\Delta C_l = C_l - C_{l_{sp}}$ on the counter-clockwise propeller blade (upper propeller of Figure 5a) in side-by-side configuration at $L_y = 2.05R$ with respect to the single rotor configuration for the last rotor revolution, $\theta = 25.5°$, $M_t = 0.32$, $J = 0.8$. At $\psi = 0°$, the tip-to-tip distance of the side-by-side blades is $0.05R$. The subscripts sp is referred to the single propeller configuration.

The polar plot of the effective angle of attack variation $\Delta\alpha_{eff}$ indicates that around $\psi = 0°$, corresponding to the azimuthal angle where the tip-to-tip distance of the side-by-side blades is $0.05R$, the blade experiences a slight increase of angle of attack at the tip region with respect to the single propeller condition (see Figure 7a). Consequently, the loads that act on the blade tip region increase along the azimuthal range of the rotor revolution, where the side-by-side propeller blades approach each other. Consequently, an increase of the sectional C_l with respect to the single propeller configuration is observed at the blade tip region around $\psi = 0°$ (see Figure 7b). On the other hand, a slight decrease of the effective angle of attack is experienced by almost all the blade sections in the range of azimuthal angle between $300° < \psi < 330°$ with a consequent decrease of the blade loading. The local behaviour of these quantities along the blade revolution reflects the fluctuations that were observed in the computed time history of the side-by side propeller thrust shown in Figure 6a.

Detailed insight regarding the flow physics involved in the side-by-side aerodynamic interaction is provided by the analysis of the propellers wake computed for the configuration characterised by the greatest interaction ($L_y = 2.05R$) when compared to the single propeller condition. In particular, Figure 8 shows the contours of the average freestream velocity component (u) calculated over the last rotor revolution on the $x - y$ plane for the single propeller and side-by-side propellers with lateral separation distance $L_y = 2.05R$.

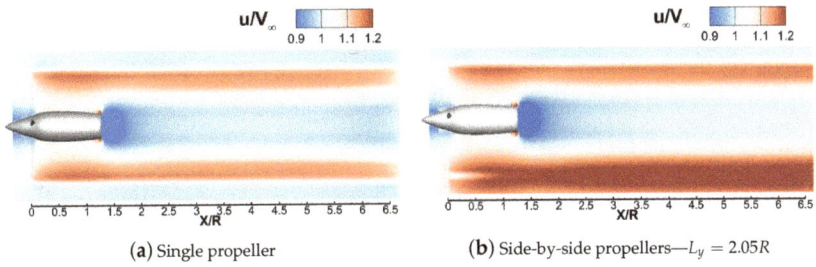

(a) Single propeller

(b) Side-by-side propellers—$L_y = 2.05R$

Figure 8. Comparison of the averaged freestream velocity component computed on the $x - y$ plane between the single propeller and side-by-side propellers configuration with $L_y = 2.05R$, $\theta = 25.5°$, $M_t = 0.32$, $J = 0.8$.

Figure 8 shows that the propeller wake in side-by-side configuration slightly expands, starting from a distance of $0.5R$ downstream the rotor disk. Consequently, a merging of the propellers wakes occurs at about $3.5R$, thus producing an increase of the resulting flow speed in this region, if compared to the single propeller case. These features were also observed in the experiments by Zhou et al. [3] performed in hover conditions, but occurring further upstream with respect to the present case. Further details of the flow physics of the side-by-side interaction are obtained analysing the instantaneous flow field calculated at $\psi = 0°$, corresponding to the azimuthal angle along rotor revolution where the interacting blades axis are aligned and the tip-to-tip distance is equal to $0.05R$. Figure 9 shows the iso-surface of the instantaneous vorticity magnitude calculated at $\psi = 0°$ for the side-by-side configuration with $L_y = 2.05$. This flow representation clearly shows the interaction between the vortices that are released by the tips of the counter-rotating propellers that leads to the merging of the iso-vorticity tubes in the region of the wake between the propellers.

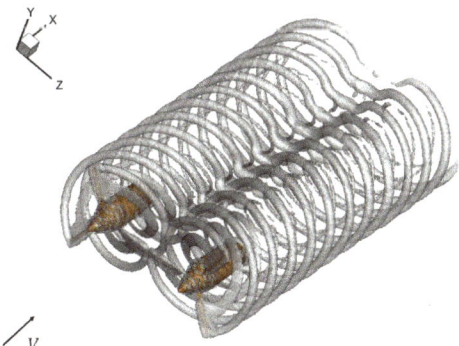

Figure 9. Iso-surface of vorticity magnitude $|\omega|$ computed for the side-by-side propellers configuration with $L_y = 2.05R$ at $\psi = 0°$, $|\omega|D/U_t = 1.1$, $\theta = 25.5°$, $M_t = 0.32$, $J = 0.8$.

A more detailed analysis of this vortex interaction can be provided by the comparison of the contours of the in-plane vorticity shown on $x - y$ plane in Figure 9. The single propeller wake is characterised by the periodic shed of counter-rotating tip vortices that are dragged downstream by the freestream velocity and conserve their relative distance showing a slow rate of dissipation throughout the entire area of investigation, as can be observed in Figure 10a. For the side-by-side interaction case, the tip vortices were found to merge starting from the distance $X/R = 0.5R$ downstream the propellers disks and dissipate much faster with respect to the single propeller case. Indeed, starting from a distance of $X/R = 1R$ downstream the propellers disks, the vortices loose their coherent structures and they are nearly unrecognizable in the region of the wake between the propellers.

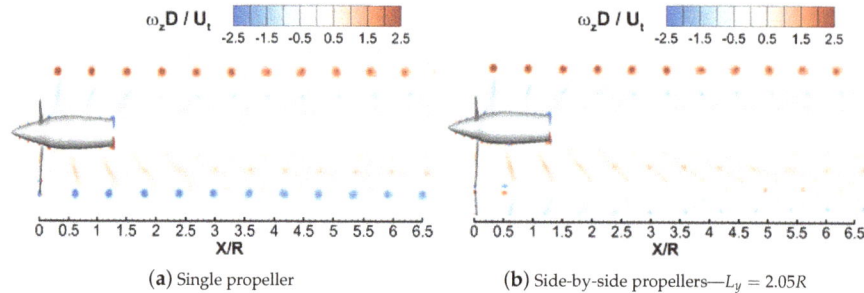

(a) Single propeller

(b) Side-by-side propellers—$L_y = 2.05R$

Figure 10. Comparison of the in-plane vorticity component ω_z computed on the $x - y$ plane between the single propeller and side-by-side propellers configuration with $L_y = 2.05R$ at $\psi = 0°$, $\theta = 25.5°$, $M_t = 0.32$, $J = 0.8$.

5.2. Tandem Propellers Configuration

In this section, the results that were obtained for the tandem propellers simulations are presented. In particular, the effect of the longitudinal distance between the propellers and the effect of the different degree of overlapping between the rotor disks is discussed by means of comparison of the computed performance coefficients and the analysis of the interacting flow field.

Figure 11 shows the time histories of the thrust coefficient C_T computed for the rear propeller (right propeller of Figure 5b) in tandem configuration with different lateral distances L_y and advance ratio $J = 0.8$. In particular, the thrust time histories are shown for both the two longitudinal distances L_x considered in the numerical activity and compared with the results that were obtained for the single propeller.

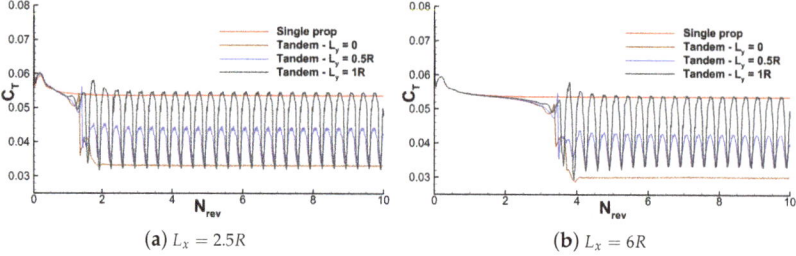

(a) $L_x = 2.5R$

(b) $L_x = 6R$

Figure 11. Time histories of the thrust coefficient C_T computed for the single propeller and for the rear propeller (right propeller of Figure 5b) in tandem configuration at different longitudinal distances L_x and lateral distances L_y, $\theta = 25.5°$, $M_t = 0.32$, $J = 0.8$.

The C_T curves behaviour clearly reflects that the rear propeller is invested by the wake of the front one at different time instances due to the different longitudinal distance between the propellers disks. In particular, the rear propeller thrust coefficient reaches a periodic behaviour after almost two and four rotor revolutions, respectively, for the longitudinal distance $L_x = 2.5R$ and $L_x = 6R$. This periodic behaviour of the computed thrust reflects the beginning of the aerodynamic interaction between the tandem propellers wakes. In particular, the aerodynamic interaction becomes stronger when the rotor disks are partially overlapped, as clearly shown by the large amplitude of the C_T fluctuations computed for $L_y = 0.5R$ and $L_y = 1R$. On the other hand, the thrust fluctuations are almost negligible when the rotor disks are completely overlapped ($L_y = 0$). As the periodicity of the loads is particularly evident over the last three revolutions of the simulations, in the following discussion the load coefficients were obtained as the averaged values calculated over this time interval.

Figure 12 shows the averaged thrust coefficient C_T, power coefficient C_P, and propulsive efficiency η of the rear propeller as function of the advance ratio J for both the considered longitudinal distances L_x and the different lateral distances L_y between the propellers disks. As can be observed from the comparison of the curves that are shown in Figure 12, the performance of the rear propeller strongly decrease when the lateral separation of the two propellers is reduced. This behaviour is apparent in the whole range of advance ratio investigated in this activity. In particular, for both the longitudinal distances L_x the highest loss of performance is obtained when the two propeller disks are completely overlapped ($L_y = 0$). By increasing the lateral distance between the propellers, the loss of the rear propeller performance decreases, as for lateral distance $L_y = 2R$ the performance curves resume the behaviour of the single rotor ones, thus confirming that, for this lateral separation, the interactional aerodynamic effects are almost negligible due to the fact that the rear rotor is unaffected by the front rotor slipstream. Moreover, the behaviour of the propulsive efficiency curves shows that the occurrence of their peaks is not affected by the tandem interaction. Indeed, for both the longitudinal distances L_x, the peak of the rear propeller propulsive efficiency η remains at $J = 0.8$ for all of the analysed lateral distances L_y.

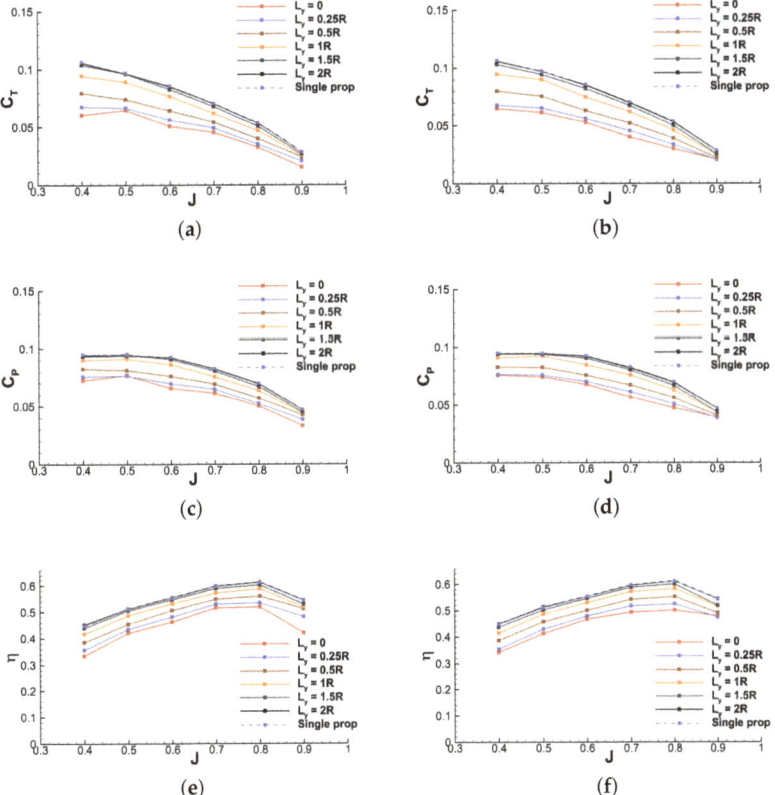

Figure 12. Results of the numerical simulations for the tandem propellers configuration. Thrust coefficient (C_T), power coefficient (C_P) and propulsive efficiency η of the rear propeller (right propeller of Figure 5b) as function of the advance ratio (J), $\theta = 25.5°$, $M_t = 0.32$; (**a,c,e**) $L_x = 2.5R$, (**b,d,f**) $L_x = 6R$.

A more detailed analysis of the rear propeller performance in tandem configuration is provided in the following for $J = 0.8$, the advance ratio corresponding to the freestream velocity considered as the target cruise velocity of urban air mobility eVTOL aircraft. In particular, Figure 13a shows the normalised thrust coefficient C_T, power coefficient C_P, and propulsive efficiency η of the rear propeller with respect to the corresponding parameters that were evaluated from the single rotor simulation. The effect of the rotor-rotor interaction on rear propeller performance is further when the degree of overlapping between the rotor disks increases, while the interactional effects become negligible when the lateral separation distance is equal to the propeller diameter ($L_y = 2R$). In particular, the higher loss of performance due to the interaction is observed when the propeller disks are completely overlapped ($L_y = 0$), where a decrease of about 45% and of more than 30%, respectively, of the thrust and power coefficients is observed for the rear propeller at $L_x = 6R$ with respect to the single propeller case. These losses lead to a reduction of about 20% of the propulsive efficiency of the rear propeller in the same tandem configuration. The effect of a lower longitudinal separation distance between the propellers is a slight decrease of the performance loss in the order of few percents of both the thrust and power coefficients of the single propeller. Indeed, for the tandem configuration with $L_x = 2.5R$, a reduction of 15% of the propulsive efficiency is observed. Moreover, Figure 13a shows that, for both the longitudinal distances between the propellers, the gradient of the performance loss is higher in the spatial range between $L_y = 0$ and $L_y = 1R$.

In order to analyse the effect of this aerodynamic interaction on the behaviour of the rear propeller loads over a rotor revolution, Figure 13b,c show the averaged C_T and C_P curves calculated at $J = 0.8$ for both the longitudinal distances L_x with errorbars representing the standard deviation of the coefficients computed over the last three revolutions. The amplitude of the loads fluctuations is higher when the lateral separation distance between the propellers is $L_y = 1R$, as can be observed from these figures. In particular, for this tandem configuration, the loads fluctuation amplitude is an order of magnitude higher than the largest value that was observed in the side-by-side configuration. The loads fluctuations level of the C_T and C_P decreases when the degree of overlapping between the propellers disks increases and become negligible when the propellers disks are completely overlapped. Moreover, for lateral separation distances higher than $L_y = 1R$, the thrust and power coefficients of the rear propeller in tandem approach the values of the single propeller configuration, while the loads fluctuations amplitude decreases.

A better insight of the effects of the aerodynamic interaction on the propeller performance already discussed is provided by Figures 14 and 15, showing the distributions of the differences of the axial velocity u_a, tangential velocity u_t, effective angle of attack α_{eff}, and sectional lift coefficient C_l of a rear propeller blade in tandem configuration with respect to the single rotor configuration computed during the last revolution. In particular, this analysis was performed for the test cases with lateral distances $L_y = 0, 0.5R, 1R$, showing the greatest interaction effects on the rear propeller aerodynamic performance, as shown in Figure 13.

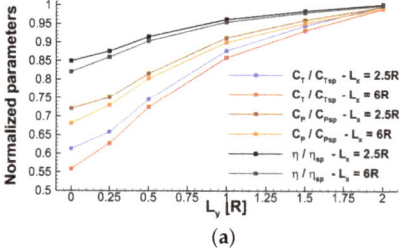

(a)

Figure 13. *Cont.*

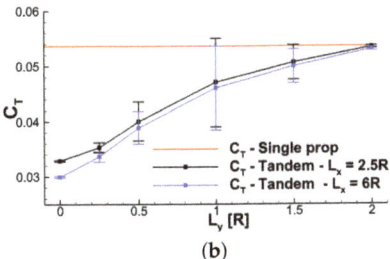
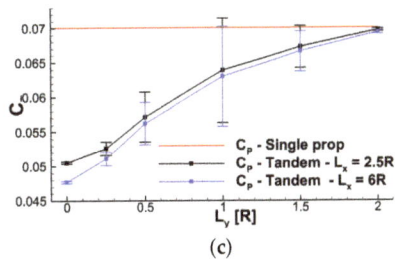

Figure 13. The results of the numerical simulations for the tandem interacting case, $\theta = 25.5°$, $M_t = 0.32$, $J = 0.8$. (**a**) Normalised thrust coefficient C_T, power coefficient C_P and propulsive efficiency η of the rear propeller with respect to the single propeller parameters as function of the lateral distance L_y. The subscript sp is referred to the single propeller configuration; (**b**,**c**) Averaged thrust coefficient C_T and power coefficient C_P of the rear propeller as function of the lateral distance L_y, errorbar corresponding to the standard deviation of the coefficients calculated over the last three rotor revolutions.

For the test configuration with the propeller disks completely overlapped ($L_y = 0$) and longitudinal distance $L_x = 2.5R$, the rear propeller blade experiences an increase of the axial velocity component with respect to freestream velocity due to the ingestion of the front propeller slipstream. This effect is particularly apparent in the outer spanwise region of the propeller blade, where the variation of the axial velocity component with respect to the single propeller reaches a value of about 5 m/s (see Figure 14a). The interaction with the front propeller slipstream also provides a slight negative variation of the tangential velocity experienced by the rear propeller blade in tandem with respect to the single propeller condition (see Figure 14d). These combined effects produce a reduction of the local effective angle of attack seen by a large region of the rear propeller blade along span (see Figure 14g). Consequently, a large reduction of the sectional lift coefficient is observed in the same spanwise region of the rear blade (see Figure 14j), thus reflecting the large loss of the average thrust calculated for this tandem configuration, as shown in Figure 13a. In particular, the axial-symmetrical behaviour that was observed for ΔC_l along propeller azimuthal angle explains the negligible amount of the loads fluctuation calculated for this tandem configuration. Similar behaviours of these variable distributions are observed for this configuration at $L_x = 6R$. In particular, due to the larger longitudinal distance between the propellers, the rear blade experiences a slightly larger increase of the axial velocity with respect to the test case with $L_x = 2.5R$ as the front propeller slipstream is further developed. This effect provides a slight higher decrease of the effective angle of attack seen by the rear propeller blade at $L_x = 6R$ and a consequent slight increase of the rear propeller performance loss with respect to the configuration with lower longitudinal distance ($L_x = 2.5R$), as shown by the average loads coefficients comparison in Figure 13.

For the test configuration with $L_y = 0.5R$, the polar plots that are shown in Figures 5 and 15 loose their axial-symmetrical behaviour, as the rear propeller disk in partially invested by the front propeller slipstream. For both the longitudinal distances tested, a conspicuous increase of the axial velocity component with respect to freestream component is observed along almost the whole blade span, particularly in the azimuthal region of the rotor revolution between $\psi = 190°$ and $\psi = 230°$.

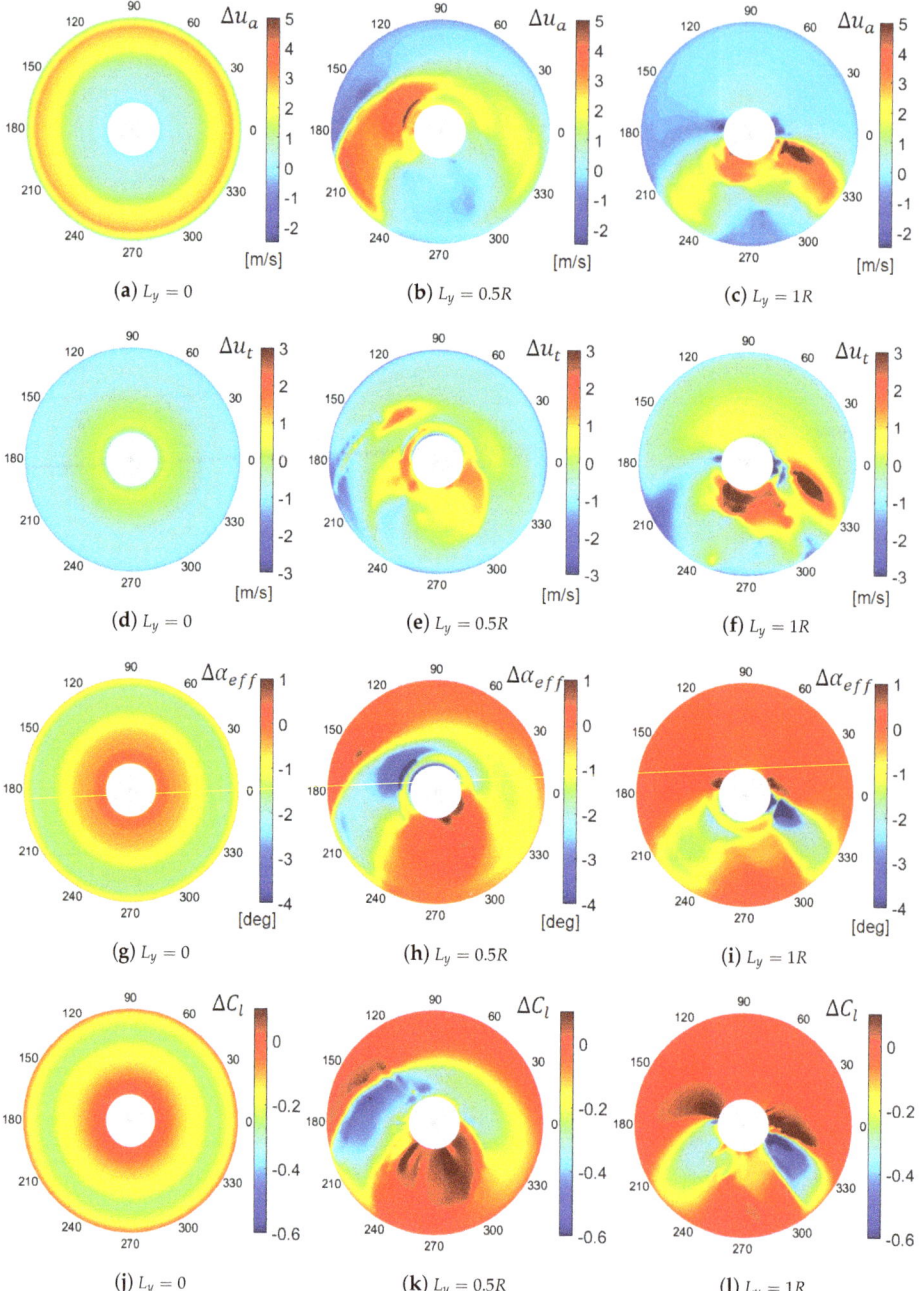

Figure 14. Variations of the axial velocity $\Delta u_a = u_a - u_{a_{sp}}$, tangential velocity $\Delta u_t = u_t - u_{t_{sp}}$, effective angle of attack $\Delta \alpha_{eff} = \alpha_{eff} - \alpha_{eff_{sp}}$ and sectional lift coefficient $\Delta C_l = C_l - C_{l_{sp}}$ on the rear propeller blade (right propeller of Figure 5b) in tandem configuration at $L_x = 2.5R$ with respect to the single rotor configuration for the last rotor revolution, $\theta = 25.5°$, $M_t = 0.32$, $J = 0.8$. The subscripts sp is referred to the single propeller configuration.

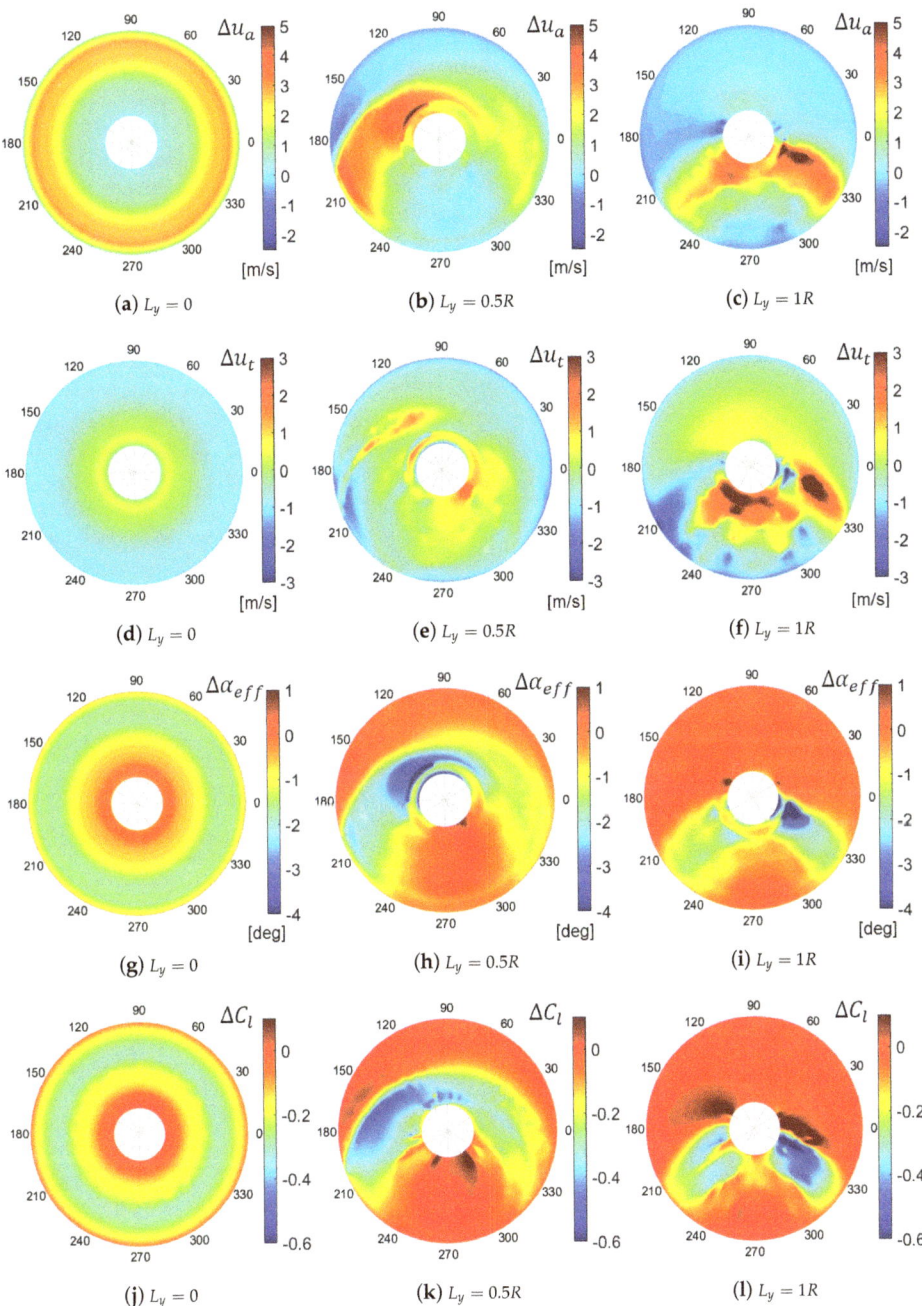

Figure 15. Variations of the axial velocity $\Delta u_a = u_a - u_{a_{sp}}$, tangential velocity $\Delta u_t = u_t - u_{t_{sp}}$, effective angle of attack $\Delta \alpha_{eff} = \alpha_{eff} - \alpha_{eff_{sp}}$ and sectional lift coefficient $\Delta C_l = C_l - C_{l_{sp}}$ on the rear propeller blade (right propeller of Figure 5b) in tandem configuration at $L_x = 6R$ with respect to the single rotor configuration for the last rotor revolution, $\theta = 25.5°$, $M_t = 0.32$, $J = 0.8$. The subscripts sp is referred to the single propeller configuration.

This effect is due to the local acceleration of the front propeller slipstream that was provided by cambered shape of the nacelle-spinner, as will be shown later in the flow visualization of Figure 19. On the other hand, a small reduction of the axial velocity is observed on the outer blade span region in the range between $\psi = 0°$ and $\psi = 180°$ (see Figures 14b and 15b), where the effect of the front propeller slipstream is negligible due to the relative position of the rotor disks. The effect of the front propeller wake interaction does not provide an apparent modification of the tangential velocity on the rear propeller blade with the exception of a smaller spanwise area of the blade that is interested by the ingestion of the outer region of the front propeller slipstream characterised by high swirl. In particular, due to the lower distance between the propellers, the rear propeller blade at $L_x = 2.5R$ experiences higher peaks of tangential velocity variations with respect to the tandem configuration with $L_x = 6R$ (see Figures 14e and 15e). The more apparent effect of the combination of the axial and tangential velocity distributions is a reduction of the effective angle of attack seen by the almost all of the rear blade sections in the azimuthal angle ranges $190° < \psi < 230°$ and $330° < \psi < 360°$ (see Figures 14h and 15h). Consequently, a quite large negative variation of the rear blade sectional lift coefficient distributions is observed in these regions with respect to the single propeller configuration (see Figures 14k and 15k), thus reflecting the remarkable amount of the loads fluctuation amplitude calculated for this tandem configuration (see Figure 13). The small difference between the average thrust losses that were calculated for the different longitudinal distances at $L_y = 0.5R$ is justified by the quite similar local behaviour observed for the spanwise sectional lift coefficient. Indeed, for this lateral distance the effect of the longitudinal distance between the propellers on the slipstream velocity components ingested by the rear propeller blade is quite small.

For the test configuration with $L_y = 1R$, the effects of the front propeller slipstream on the rear propeller blade is smaller with respect to the previous analysed configuration due to the lower degree of overlapping between the propellers disks in tandem. In particular, the most evident effects of this interaction is observed in the azimuthal angle range of the rotor revolution between $\psi = 180°$ and $\psi = 360°$. For both of the longitudinal distances tested, the axial velocity component behaviour shows a concentrated increase in this azimuthal angle range due to the local acceleration of the front propeller slipstream in this area provided by the curvature of the nacelle-spinner surface (see Figures 14c and 15c), as will be shown later in the flow visualization of Figure 22. In the same region, the interaction of the front propeller slipstream provides a large increase of the tangential velocity (see Figures 14f and 15f), thus the combination of these velocity components variations provides a remarkable decrease of the effective angle of attack seen by the rear blade sections in the ranges $210° < \psi < 240°$ and $300° < \psi < 330°$ (see Figures 14i and 15i) and a consequent high variation of the sectional lift in the same areas. The behaviour of the sectional lift variation reflects the larger amplitude of the loads fluctuations observed for this tandem configuration (see Figure 13). Additionally, for this lateral distance, a quite low effect due to the longitudinal distance is observed from the representations of the analysed local quantities reflecting the results comparison in terms of the average rear propeller performance.

Global insight regarding the flow physics that are involved in the aerodynamic interaction for the investigated tandem configurations is provided by the following analysis of the propellers wakes computed at $J = 0.8$ for the same lateral distances $L_y = 0, 0.5R, 1R$ discussed before. For the test configuration with the propeller disks completely overlapped ($L_y = 0$), Figure 16 shows the contours of the average freestream velocity component (u) calculated over the last rotor revolution on the $x - y$ plane. The averaged flow fields clearly show that, for both the longitudinal distances L_x, the wake of the rear propeller is quite faster at the tip region of the rotor disk with respect to the front propeller one. Indeed, the co-axial configuration of the two propellers provides a combination of the accelerated flow regions passing through the outer regions of the propeller disks.

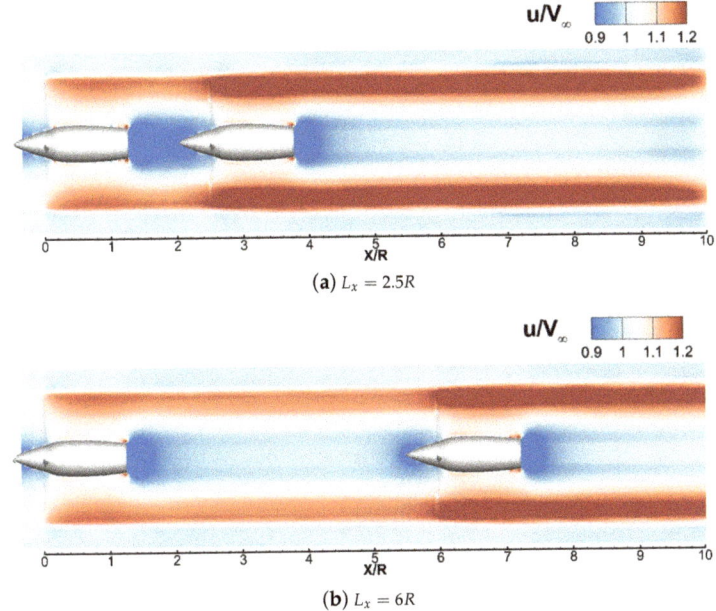

Figure 16. Comparison of the averaged freestream velocity component computed on the $x-y$ plane for the tandem propellers configurations with $L_y = 0$, $\theta = 25.5°$, $M_t = 0.32$, $J = 0.8$.

A more detailed understanding of the flow physics that are involved in the wakes interaction is obtained by the instantaneous flow field shown for the azimuthal angle $\psi = 0°$ in Figure 17 by means of the iso-surfaces of the instantaneous vorticity magnitude. This flow representation clearly shows that, for both of the considered distances L_x, the interaction between the propellers wakes does not affect the coherence of the helical structure of vorticity released by the rear propeller. In particular, an increase of the iso-vorticity tubes is observed downstream the rear propeller due to the coalescence of the vortical structures that are released by the two co-axial propellers in tandem.

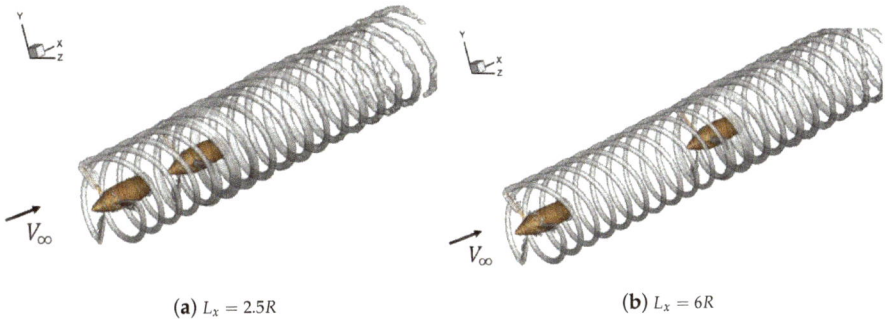

Figure 17. Iso-surface of vorticity magnitude $|\omega|$ computed for the tandem propellers configuration with $L_y = 0$ at $\psi = 0°$, $|\omega| D/U_t = 1.45$, $\theta = 25.5°$, $M_t = 0.32$, $J = 0.8$.

Figure 18 shows more details of the aerodynamic interaction by means of the contours of the in-plane vorticity calculated on the $x-y$ from the instantaneous flow field at $\psi = 0°$.

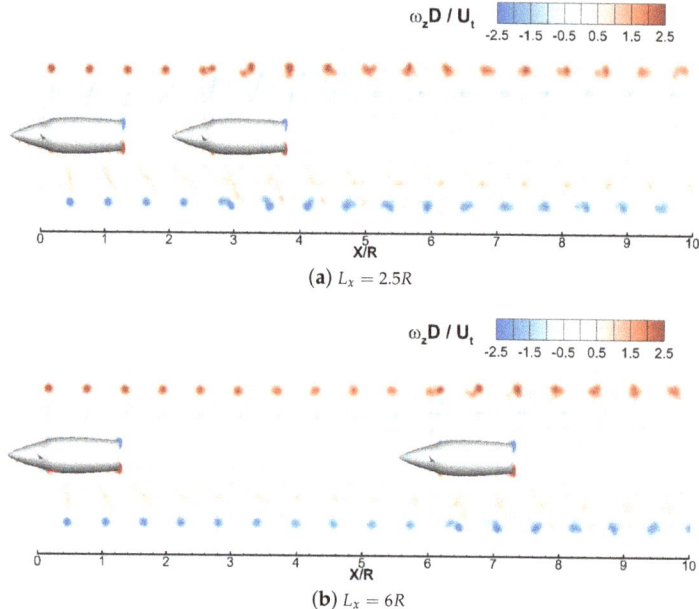

Figure 18. Comparison of the in-plane vorticity component ω_z computed on the $x - y$ plane for the tandem propellers configurations with $L_y = 0$ at $\psi = 0°$, $\theta = 25.5°$, $M_t = 0.32$, $J = 0.8$.

Because, in the present simulations, the blades of the propellers in tandem are co-rotating and synchronised, the tip vortices shed by the front propeller blades interact with the ones released by the rear propeller providing downstream the rotor disk co-rotating vortical structures that are characterised by higher vorticity and larger core. This feature is observed for both the longitudinal distances analysed, but for $L_x = 2.5R$ the resulting vortical structures show a higher level of vorticity with respect to the configuration with $L_x = 6R$. Indeed, the vortices released by the front propeller blades are less dissipated when they interact with the ones released by the rear propeller, due to the lower longitudinal distance between the propellers. This feature is highlighted by the higher intensity of red that characterise the representation of the vortex investing the tip region of the rear propeller blade for $L_x = 2.5R$ with respect to the test configuration with $L_x = 6R$ (see Figure 18a,b).

For the tandem configuration with lateral distance $L_y = 0.5$, the averaged flow fields that are presented in Figure 19 show an asymmetrical behaviour of the rear propeller wake with respect to the longitudinal axis due to the interaction with the front propeller slipstream. Indeed, for this configuration the rear propeller disk is only partially invested by the front propeller slipstream, thus the lower region of the rear propeller wake is accelerated by the effect of the front propeller wake. This effect is particularly evident for the lower longitudinal distance $L_x = 2.5R$. Moreover, the upper region of the front propeller slipstream is dragged upward and locally accelerated by the the cambered shape of the nacelle-spinner surface.

The three-dimensional representation of the instantaneous flow field provided by the iso-surfaces of vorticity in Figure 20 shows, for this configuration ($L_y = 0.5R$), a stronger interaction between the vortical structures released the propeller blades with respect to the test case where the propeller disks are completely overlapped ($L_y = 0$). Indeed, for both the longitudinal distances between the propellers the helical structures of vorticity released by the rear propeller blades loose their coherence due to the interaction.

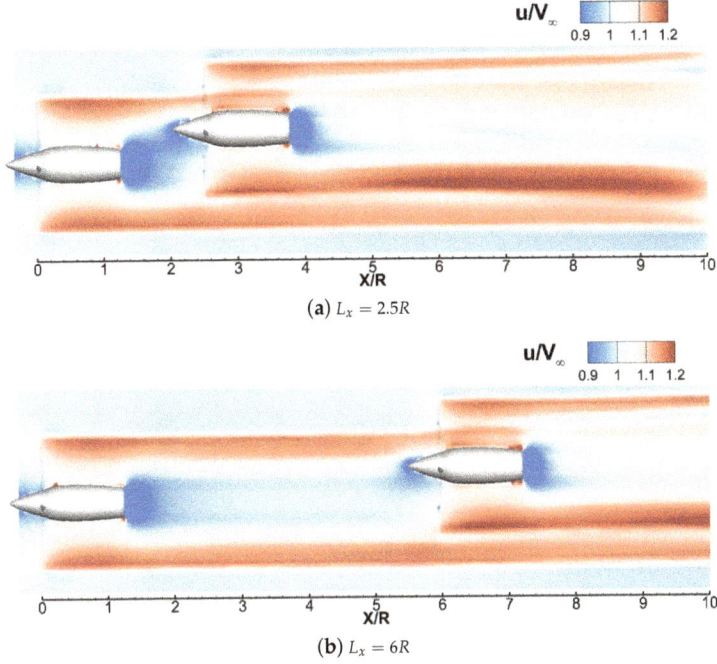

Figure 19. Comparison of the averaged freestream velocity component computed on the $x - y$ plane for the tandem propellers configurations with $L_y = 0.5R$, $\theta = 25.5°$, $M_t = 0.32$, $J = 0.8$.

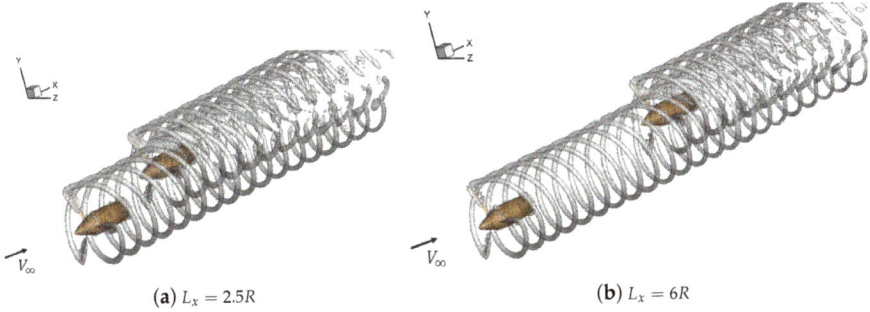

Figure 20. Iso-surface of vorticity magnitude $|\omega|$ computed for the tandem propellers configuration with $L_y = 0.5R$ at $\psi = 0°$, $|\omega|D/U_t = 1.45$, $\theta = 25.5°$, $M_t = 0.32$, $J = 0.8$.

More details regarding the interaction of the propellers wakes is provided by the representation of the in-plane vorticity contours that are shown in Figure 21. A relevant effect of the interaction is that the upper region of the front propeller wake diverges upward due to the presence of the rear propeller nacelle. Therefore, in the upper region past, the rear propeller disk the vortices that are released by the front propeller blades are dragged toward the ones released by the rear propeller, thus producing a pairing of the co-rotating vortices that provides the winding of the shear layer into a series of counter-rotating vortices. This feature is observed for both the longitudinal distances of the propellers in tandem.

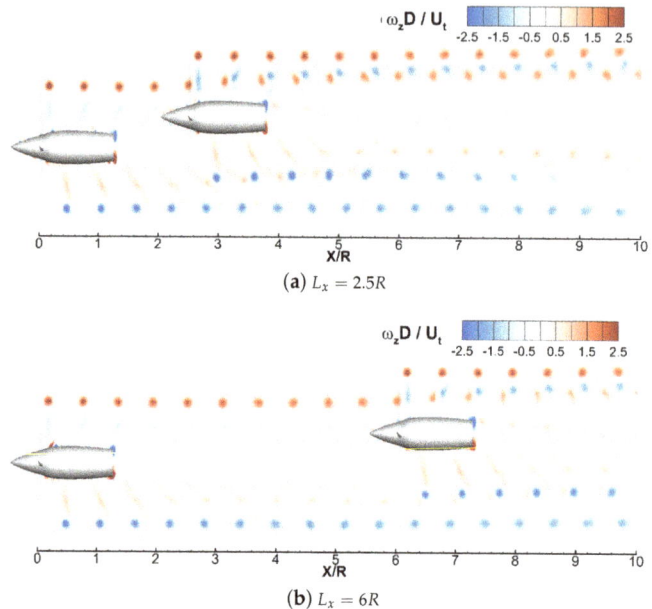

Figure 21. Comparison of the in-plane vorticity component ω_z computed on the $x - y$ plane for the tandem propellers configurations with $L_y = 0.5R$ at $\psi = 0°$, $\theta = 25.5°$, $M_t = 0.32$, $J = 0.8$.

For the tandem configuration with lateral distance $L_y = 1R$, the averaged flow field presented in Figure 22 shows an asymmetrical behaviour of the rear propeller wake similarly to what was found for $L_y = 0.5R$. In particular, for this test case, due to the higher degree of overlapping between the propeller disks, the upper region of the front propeller slipstream is dragged downward and locally accelerated by the presence of the rear propeller nacelle. Thus, an increase of the area of accelerated flow can be observed in the lower region of the rear propeller wake. This effect is more pronounced for the tandem configuration with $L_x = 2.5R$.

The instantaneous flow representation that is provided by the iso-surfaces of vorticity shows, similarly to the test case with $L_y = 0.5R$, that the wakes interaction produces a break of the coherent helical structures released by the rear propeller blades (see Figure 23. In particular, the in-plane vorticity field presented in Figure 24 shows that for this lateral distance ($L_y = 1R$) the tip vortices shed by the front propeller blades dissipate once they impinge the rear propeller nacelle nose. Indeed, the trace of these vortices is negligible downstream the rear propeller disk.

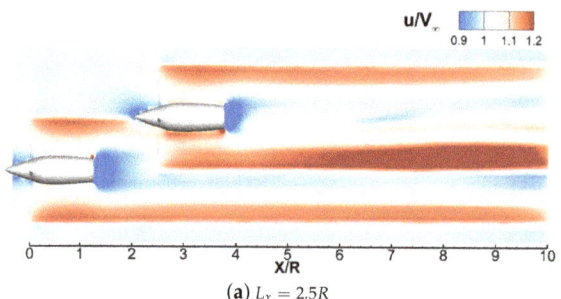

(a) $L_x = 2.5R$

Figure 22. *Cont.*

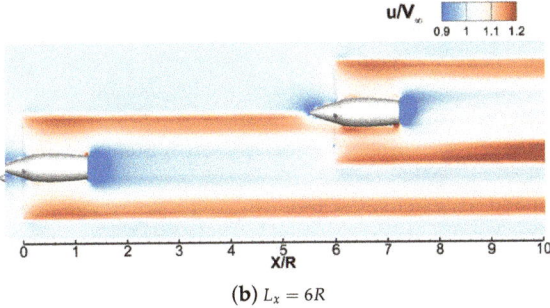

(b) $L_x = 6R$

Figure 22. Comparison of the averaged freestream velocity component computed on the $x - y$ plane for the tandem propellers configurations with $L_y = 1R$, $\theta = 25.5°$, $M_t = 0.32$, $J = 0.8$.

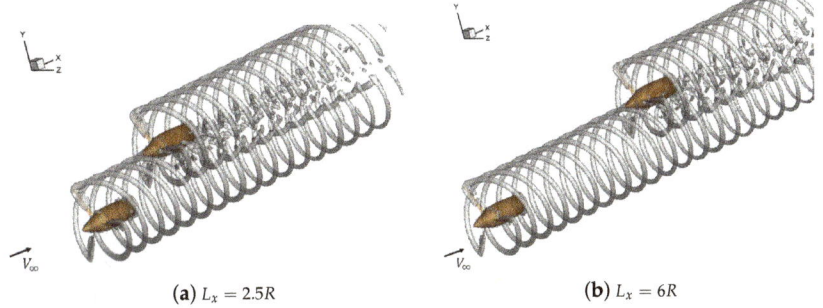

(a) $L_x = 2.5R$ (b) $L_x = 6R$

Figure 23. Iso-surface of vorticity magnitude $|\omega|$ computed for the tandem propellers configuration with $L_y = 1R$ at $\psi = 0°$, $|\omega|D/U_t = 1.45$, $\theta = 25.5°$, $M_t = 0.32$, $J = 0.8$.

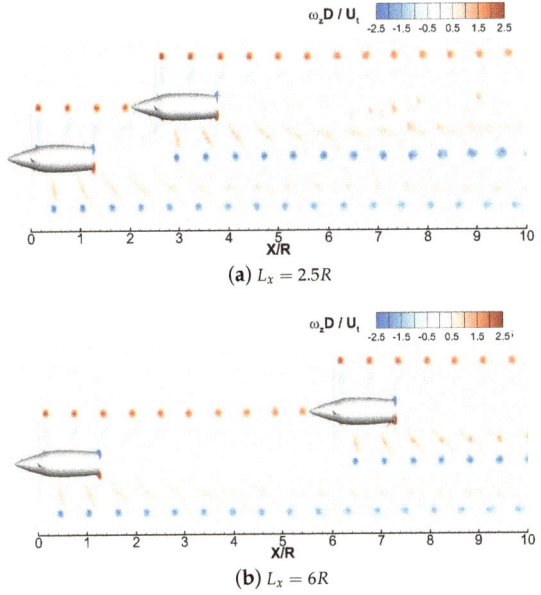

Figure 24. Comparison of the in-plane vorticity component ω_z computed on the mid-span $x - y$ plane for the tandem propellers configurations with $L_y = 1R$ at $\psi = 0°$, $\theta = 25.5°$, $M_t = 0.32$, $J = 0.8$.

6. Conclusions

A numerical activity was performed in order to investigate the rotor-rotor aerodynamic interaction that is typical of the novel architectures of eVTOL aircraft designed for urban air mobility. With this aim, a systematic study of the aerodynamic interaction of two propellers both in side-side and tandem configurations was performed while using a mid-fidelity aerodynamic solver. The low computational effort required by the solver enabled to simulate a comprehensive set of propellers configurations at different advance ratios, thus providing a comprehensive numerical database to eVTOL research community to be used to drive the design of new unconventional aircraft configurations. In particular, the numerical simulations enabled investigating the effects of these kinds of aerodynamic interactions both on propellers performance and flow physics involved. The discussion was particularly focused on the results that were obtained for the propellers advance ratio corresponding to the target cruise velocity of a eVTOL aircraft in urban areas.

The main results that were obtained from the side-by-side simulations showed that the greatest effect of the aerodynamic interaction on the propellers performance is reached at the lowest lateral separation distance between them. In particular, a slight reduction of the average propeller thrust and propulsive efficiency below 1% was found for the interacting case with respect to the single propeller configuration. On the other hand, a high amplitude of load fluctuations is observed for this test condition that could provide a drawback for aeroacoustic issues. Moreover, the visualizations of the instantaneous flow field illustrate the strong interaction between the tip vortices that were released by the two side propellers.

A more comprehensive investigation was performed for the tandem configuration that aimed to evaluate the different effect of the aerodynamic interaction due to the longitudinal and lateral distance between the propellers at several advance ratios. The results analysis, focused on the advance ratio corresponding to the target cruise speed of eVTOLs, showed that a remarkable decrease of the rear propeller performance is observed due to this aerodynamic interaction. In particular, a loss of the average thrust of the rear propeller in tandem in the order of 40% with respect to the single propeller and a reduction of about 20% of the propulsive efficiency was found when the propellers disks are completely overlapped. In particular, the effect of a lower longitudinal distance between the propellers is a slight reduction of the performance losses on the rear propeller. The performance losses that were evaluated on the rear propeller in tandem were discussed, analysing the axial and tangential velocity components of the slipstream investing the rear propeller blade during a rotor revolution and the consequent distributions of the effective angle of attack seen by the blade propeller sections and the sectional loads. The analysis of the local loads acting on the rear propeller blade shows that a partial overlapping between the propellers in tandem provides a lower effect on the average loads, but a larger amplitude of loads fluctuation along a rotor revolution with respect to the co-axial configuration. Moreover, a deeper insight regarding the flow physics that are involved in the interaction between the wakes of the two propellers in tandem was provided by means of the analysis of the averaged and instantaneous flow fields for the three lateral separation distances between the propellers characterised by the highest effects on the rear propeller performance. The flow fields analysis illustrated how the front propeller slipstream interacts with the wake of the rear propeller, showing, in particular, the pairing between the tip vortices that are released by the two propellers that occur when the rotor disks are co-axial or present a low degree of overlapping.

Author Contributions: Conceptualization, A.Z.; methodology, R.P., A.Z.; software, M.T.; validation, R.P., M.T.; formal analysis, R.P.; investigation, R.P.; resources, R.P.; data curation, R.P., A.Z.; writing—original draft preparation, A.Z.; writing—review and editing, A.Z.; visualization, A.Z.; supervision, A.Z.; project administration, A.Z.; funding acquisition, A.Z. All authors have read and agreed to the published version of the manuscript.

Funding: This research received no external funding.

Conflicts of Interest: The authors declare no conflict of interest.

Abbreviations

The following nomenclature and abbreviations are used in this manuscript:

CFD	Computational Fluid Dynamics
C_l	sectional lift coefficient
C_P	power coefficient = $P/(\rho n^3 D^5)$
C_Q	power coefficient = $Q/(\rho n^2 D^5)$
C_T	thrust coefficient = $T/(\rho n^2 D^4)$
D	propeller diameter [m]
eVTOL	electrical Vertical Take Off and Landing aircraft
J	advance ratio = $V_\infty/(nD)$
L_x	longitudinal distance between the propeller disks [m]
L_y	lateral distance between the propeller axis [m]
M_t	tip Mach number
n	rotational speed [rad/s]
N_rev	number of rotor revolutions
P	propeller power [W]
Q	propeller torque [Nm]
R	rotor radius [m]
Re_D	Reynolds number based on propeller diameter
T	propeller thrust [N]
u	freestream velocity component [m/s]
u_a	blade axial velocity component [m/s]
u_t	blade tangential velocity component [m/s]
U_t	velocity at blade tip [m/s]
VPM	Vortex Particle Method
V_∞	freestream velocity [m/s]
α_{eff}	effective angle of attack [deg]
η	propulsive efficiency = $J(C_T/C_P)$
ψ	blade azimuthal angle [deg]
ρ	air density [kg/m^3]
θ	blade pitch angle at 75% of the rotor radius [deg]
$|\omega|$	vorticity magnitude [1/s]
ω_x	in-plane vorticity component [1/s]

References

1. Polaczyk, N.; Trombino, E.; Wei, P.; Mitici, M. A Review of Current Technology and Research in Urban On-Demand Air Mobility Applications. In Proceedings of the 8th Biennial Autonomous VTOL Technical Meeting and 6th Annual Electric VTOL Symposium, Mesa, AZ, USA, 29–31 January 2019.
2. Droandi, G.; Syal, M.; Bower, G. Tiltwing Multi-Rotor Aerodynamic Modeling in Hover, Transition and Cruise Flight Conditions. In Proceedings of the 74th Annual American Helicopter Society International Forum and Technology Display 2018 (FORUM 74), Phoenix, AZ, USA, 14–17 May 2018.
3. Zhou, W.; Ning, Z.; Li, H.; Hu, H. An Experimental Investigation on Rotor-to-Rotor Interactions of Small UAV Propellers. In Proceedings of the 35th AIAA Applied Aerodynamics Conference, Denver, CO, USA, 5–9 June 2017.
4. Shukla, D.; Komerath, N. Multirotor Drone Aerodynamic Interaction Investigation. *Drones* **2018**, *2*, 43. [CrossRef]
5. Shukla, D.; Hiremath, N.; Komerath, N.M. Low Reynolds Number Aerodynamics Study on Coaxial and Quad-Rotor. In Proceedings of the 53th AIAA Aviation Forum, Atlanta, GA, USA, 25–29 June 2018.
6. Brazinskas, M.; Prior, S.; Scanlan, J. An Empirical Study of Overlapping Rotor Interference for a Small Unmanned Aircraft Propulsion System. *Aerospace* **2016**, *3*, 32. [CrossRef]
7. Yoon, S.; Lee, H.; Pulliam, T. Computational Analysis of Multi-Rotor Flows. In Proceedings of the 54th AIAA Aerospace Sciences Meeting, San Diego, CA, USA, 4–8 January 2016.

8. Yin, J.; Ahmed, S. Helicopter Main-Rotor/Tail-Rotor Interaction. *J. Am. Helicopter Soc.* **2000**, *4*, 293–302. [CrossRef]
9. Wentrup, M.; Yin, J.; Kunze, P.; Streit, T.; Wendisch, J.; Schwarz, T.; Pinacho, J.; Kicker, K.; Fukari, R. An overview of DLR compound rotorcraft aerodynamics and aeroacoustics activities within the CleanSky2 NACOR Project. In Proceedings of the 74th AHS Annual Forum & Technology Display, Phoenix, AZ, USA, 14–17 May 2018.
10. Cottet, G.H.; Koumoutsakos, P.D.; Petros, D. *Vortex Methods: Theory and Practice*; Cambridge University Press: Cambridge, UK, 2000.
11. Winckelmans, G.S. Topics in Vortex Methods for the Computation of Three-and Two-Dimensional Incompressible Unsteady Flows. Ph.D. Thesis, California Institute of Technology, Pasadena, CA, USA, 1989.
12. Su, T.; Lu, Y.; Ma, J.; Guan, S. Aerodynamic characteristics analysis of electrically controlled rotor based on viscous vortex particle method. *Aerosp. Sci. Technol.* **2020**, *97*, 105645. [CrossRef]
13. Lu, Y.; Su, T.; Chen, R.; Li, P.; Wang, Y. A method for optimizing the aerodynamic layout of a helicopter that reduces the effects of aerodynamic interaction. *Aerosp. Sci. Technol.* **2019**, *88*, 73–83. [CrossRef]
14. Tan, J.; Sun, J.; Barakos, G. Unsteady loads for coaxial rotors in forward flight computed using a vortex particle method. *Aeronaut. J.* **2018**, *122*, 693–714. [CrossRef]
15. Tan, J.; Zhou, T.; Sun, J.; Barakos, G. Numerical investigation of the aerodynamic interaction between a tiltrotor and a tandem rotor during shipboard operations. *Aerosp. Sci. Technol.* **2019**, *87*, 62–72. [CrossRef]
16. Alvarez, E.; Ning, A. Modeling Multirotor Aerodynamic Interactions Through the Vortex Particle Method. In Proceedings of the 54th AIAA Aviation Forum, Dallas, TX, USA, 17–21 June 2019.
17. Montagnani, D.; Tugnoli, M.; Fonte, F.; Zanotti, A.; Droandi, G.; Syal, M. Mid-Fidelity Analysis of Unsteady Interactional Aerodynamics of Complex VTOL Configurations. In Proceedings of the 45th European Rotorcraft Forum, Warsaw, Poland, 17–20 September 2019.
18. Droandi, G.; Syal, M.; Bower, G. Analysis of the Interactional Aerodynamics of the Vahana eVTOL Using a Medium Fidelity Open Source Tool. In Proceedings of the VFS Aeromechanics for Advanced Vertical Flight Technical Meeting, San Jose, CA, USA, 21–23 January 2020.
19. Lindsay, K.; Krasny, R. A particle method and adaptive treecode for vortex sheet motion in three-dimensional flow. *J. Comput. Phys.* **2001**, *172*, 879–907. [CrossRef]
20. Morino, L.; Kuot, C.C. Subsonic potential aerodynamics for complex configurations: A general theory. *AIAA J.* **1974**, *12*, 191–197. [CrossRef]
21. Gallay, S.; Laurendeau, E. Nonlinear generalized lifting-line coupling algorithms for pre/poststall flows. *AIAA J.* **2015**, *53*, 1784–1792. [CrossRef]
22. Piszkin, S.T.; Levinsky, E. *Nonlinear Lifting Line Theory for Predicting Stalling Instabilities on Wings of Moderate Aspect Ratio*; Technical Report; General Dynamics Convair Division Co.: San Diego, CA, USA, 1976.
23. Brown, R.E.; Line, A.J. Efficient high-resolution wake modeling using the vorticity transport equation. *AIAA J.* **2005**, *43*, 1434–1443. [CrossRef]
24. McCrink, M.H.; Gregory, J.W. Blade Element Momentum Modeling of Low-Reynolds Electric Propulsion Systems. *J. Aircr.* **2017**, *54*, 163–176. [CrossRef]
25. Drela, M. XFOIL: An Analysis and Design System for Low Reynolds Number Airfoils. In *Low Reynolds Number Aerodynamics*; Mueller, T.J., Ed.; Springer: Berlin/Heidelberg, Germany, 1989; pp. 1–12.
26. Viterna, L.A.; Janetzke, D.C. *Theoretical and Experimental Power from Large Horizontal-Axis Wind Turbines*; Technical Report; Washington Procurement Operations Office: Washington, DC, USA, 1982.
27. Piccinini, R. Rotor-Rotor Aerodynamic Interactions for eVTOL Aircraft Configurations. Master's Thesis, Politecnico di Milano, Milan, Italy, 2020.

Publisher's Note: MDPI stays neutral with regard to jurisdictional claims in published maps and institutional affiliations.

© 2020 by the authors. Licensee MDPI, Basel, Switzerland. This article is an open access article distributed under the terms and conditions of the Creative Commons Attribution (CC BY) license (http://creativecommons.org/licenses/by/4.0/).

Article

Coupling Mid-Fidelity Aerodynamics and Multibody Dynamics for the Aeroelastic Analysis of Rotary-Wing Vehicles

Alberto Savino, Alessandro Cocco, Alex Zanotti, Matteo Tugnoli, Pierangelo Masarati and Vincenzo Muscarello

Dipartimento di Scienze e Tecnologie Aerospaziali, Politecnico di Milano, Via La Masa 34, 20156 Milan, Italy; alberto.savino@polimi.it (A.S.); alessandro.cocco@polimi.it (A.C.); matteo.tugnoli@polimi.it (M.T.); pierangelo.masarati@polimi.it (P.M.); vincenzo.muscarello@polimi.it (V.M.)
* Correspondence: alex.zanotti@polimi.it

Abstract: A mid-fidelity aerodynamic solver based on the vortex particle method for wake modeling, DUST, is coupled through the partitioned multi-physics coupling library preCICE to a multibody dynamics code, MBDyn, to improve the accuracy of aeroelastic numerical analysis performed on rotary-wing vehicles. In this paper, the coupled tool is firstly validated by solving simple fixed-wing and rotary-wing problems from the open literature. The transient roll maneuver of a complete tiltrotor aircraft is then simulated, to show the capability of the coupled solver to analyze the aeroelasticity of complex rotorcraft configurations. Simulation results show the importance of the accurate representation of rotary wing aerodynamics provided by the vortex particle method for loads evaluation, aeroelastic stability assessment, and analysis of transient maneuvers of aircraft configurations characterized by complex interactional aerodynamics. The limited computational effort required by the mid-fidelity aerodynamic approach represents an effective trade-off in obtaining fast and accurate solutions that can be used for the preliminary design of novel rotary-wing vehicle configurations.

Citation: Savino, A.; Cocco, A.; Zanotti, A.; Tugnoli, M.; Masarati, P.; Muscarello, V. Coupling Mid-Fidelity Aerodynamics and Multibody Dynamics for the Aeroelastic Analysis of Rotary-Wing Vehicles. *Energies* **2021**, *14*, 6979. https://doi.org/10.3390/en14216979

Academic Editor: Chunhua Liu

Received: 20 September 2021
Accepted: 19 October 2021
Published: 25 October 2021

Publisher's Note: MDPI stays neutral with regard to jurisdictional claims in published maps and institutional affiliations.

Copyright: © 2021 by the authors. Licensee MDPI, Basel, Switzerland. This article is an open access article distributed under the terms and conditions of the Creative Commons Attribution (CC BY) license (https://creativecommons.org/licenses/by/4.0/).

Keywords: aeroelasticity; fluid-structure interaction; rotary-wing aerodynamics; multibody dynamics; tiltrotor; computational fluid dynamics; vortex particle method

1. Introduction

The design of complex rotary-wing vehicles, such as tiltrotors, represents a challenge for engineers and scientists. Being able to perform complete aeroelastic simulations considering the coupling of rotary-wing aerodynamics with structural dynamics of the vehicle is essential for the development of novel aircraft configurations [1]. Sophisticated numerical tools have been developed to support the analysis of rather different operating flight conditions typical of VTOL vehicles with increasing level of detail. Indeed, a substantial effort was spent in past years to develop sophisticated structural dynamics codes (CSD) that were effectively used for rotorcraft applications (e.g., References [2–4]). In detail, structural dynamics of rotary-wing vehicles was typically investigated using the multibody approach [5,6], which takes into account the nonlinear dynamics of the interconnected rigid and flexible bodies representing the aircraft components during transients. The multibody approach was also used to investigate aeroelastic phenomena, especially in airplane mode flight where whirl-flutter instabilities may occur [7]. A particular effort in this research field was spent at Politecnico di Milano, where, starting in the 1990s, a free general-purpose multibody software called MBDyn (http://www.mbdyn.org/) was developed, with the aim of gaining autonomous modeling capabilities of generic problems related to the dynamics of complex aeroelastic systems, specifically rotorcraft and tiltrotors [8]. A tiltrotor flight mission is characterized by vertical take off and landing, typical of helicopters, and by a cruise flight condition typical of fixed-wing airplanes. Transition between the two flight conditions occurs through the so-called conversion maneuver, in which the proprotor-nacelle system is tilted. Thus, an accurate model for the evaluation of unsteady loads occurring during transients must be considered for a correct analysis of the complex dynamics characterizing of these vehicles. However, multibody solvers, such as MBDyn, are typically equipped with simplified aerodynamics models based on Blade Element/Momentum Theory (BE/MT). This type of approach does not consider aerodynamic interference between rotors and the actual geometry of lifting surfaces, possibly leading to a misrepresentation of the aerodynamic loads and loss of information related to periodic actions.

The accurate computation of unsteady aerodynamic loads for rotary-wing vehicles was the main goal of several research groups that developed, in recent decades, different high fidelity Computational Fluid Dynamics (CFD) codes employing the use of high-fidelity Navier–Stokes equations solvers. To name a few examples, ONERA, University of Glasgow, DLR and Airbus Helicopters Deutschland and Politecnico di Milano developed, respectively, elsA [9], HBM3 [10], FLOWer [11], and ROSITA [12], high-fidelity CFD codes based on the block-structured grid, finite volume, and Chimera approach for the simulation of rotating bodies. These tools were purposely developed in Europe for rotorcraft and tiltrotor application studies [13–15]. Similarly, considerable research effort was dedicated in the USA to the numerical study of rotorcraft, particularly to tiltrotor aerodynamics, as shown for instance by the works by Meakin [16], Potsdam and Strawn [17], and Wissink et al. [18], where hover configurations were simulated using different implementations of the Navier–Stokes equations, and by the recent works of Lim, Tran et al. [19–22] that investigated the aerodynamic interaction between rotors and wing of the XV-15 tiltrotor aircraft.

Coupling of CSD codes with high fidelity CFD solvers was successfully investigated and implemented in the last two decades years for aeroelastic simulation of rotorcraft applications [23–28]. The coupled CSD/CFD numerical approach was successfully validated against experimental results, e.g., for the flutter calculations of a vertical tail model [29] and for the analysis of rotor blade structural loads of a complete helicopter model tested in a transonic wind tunnel [30]. Nevertheless, despite continuous advances in the field of high performance computing, coupled simulations of CSD and time-accurate URANS simulations of complete rotorcraft configurations still require a robust computational effort, not suitable for the preliminary design stage of novel VTOL aircraft configurations as tiltrotors, which requires a great number of simulations to reproduce the different flight conditions that characterize their mission.

In recent years, considerable effort was dedicated by several research groups to the development of mid-fidelity aerodynamic solvers based on the use of the vortex particle method (VPM) [31,32] for wake modeling. This numerical methodology showed a quite accurate representation of the aerodynamic interactions among several bodies typical of complex rotorcraft configurations and limited computational time with respect to URANS CFD simulations. To cite few examples, Lu et al. [33] developed an optimization methodology for helicopter design based on an viscous VPM model combined with an unsteady panel hybrid method. Alvarez and Ning developed a VPM-based code [34] for the investigation of multi-rotor configurations. Tan et al. [35] used a vortex-based approach coupled with a viscous boundary model to study rotor-to-rotor interactional problems occurring during shipboard operations [35]. Recently, Politecnico di Milano developed a novel, flexible mid-fidelity computational tool, called DUST (www.dust-project.org) aimed at representing a fast and reliable asset for the simulation of the aerodynamics of complex rotorcraft configurations, such as the electrical Vertical Take Off and Landing (eVTOL) aircraft. DUST is an open source code, released under MIT license, integrating different aerodynamic models for solid bodies, as thick surface panels, thin vortex lattices elements, and lifting lines elements. Moreover, a VPM method was implemented for wake modeling providing a stable Lagrangian description of free-vorticity flow field, which is suitable for numerical simulations of configurations characterized by strong aerodynamic interactions. DUST was thoroughly validated by comparison with both high-fidelity CFD simulations results and experimental data over complex rotorcraft configurations, such as eVTOLs [36] and tiltrotors [37]. Consequently, this novel open source tool is reaching maturity for the simulation of the aerodynamics of complex rotorcraft configurations accurately reproducing the interaction between rotors and wings.

The combination of a multibody dynamics solver with a mid-fidelity aerodynamic tool aims at representing an ideal trade-off between speed of execution and accuracy of the solution, aimed at the preliminary design of novel rotary-wing aircraft configurations. A novel open access aeroelastic tool was built by coupling MBDyn with DUST. The coupling of the two codes relies on the partitioned multi-physics coupling library preCICE [38], a very useful and robust tool for managing the communication between different solvers. After a brief description of the multibody dynamics and mid-fidelity aerodynamics solvers, this paper describes the details of the methodology used for the coupling. An interesting novelty proposed by this tool is the capability of modeling the deflection of a control surface, representing an essential aspect in the simulation of aircraft maneuvers. The coupled code is validated considering two test cases [39,40]. The first validation is provided by analyzing Goland's wing, a numerical test case that was widely used in literature as a benchmark for flutter predictions. The second test case involves a rotary-wing application, using experimental data available for the XV-15 tiltrotor in hover. Then, the paper shows the capability of the tool to simulate a more complex application, namely the coupled simulation of a complete tiltrotor during a transient maneuver, a novel application for a coupled CSD/mid-fidelity aerodynamic tool in the rotorcraft field. Indeed, the control of these vehicles is often obtained by concurrently operating several control surfaces and actuation systems associated with the rotors, mixed by a complex Flight Control System (FCS) that takes into account the different flight conditions [41,42]. Thus, the design of the control surfaces and the selection of the actuators requires the ability to evaluate the unsteady aeroelastic loads acting on the aircraft during the maneuvers, to improve the vehicle response, increase effectiveness and efficiency, and reduce weight and complexity of the control system. Consequently, the coupled simulation of the complete tiltrotor model, representative of the Bell XV-15, was aimed at indicating the role of an accurate reproduction of rotary-wing aerodynamics and aeroelasticity, resulting from VPM and including the effects of the mutual interactions between the components of the complete aircraft, to correctly reproduce the dynamics of a roll maneuver. The results represent a key-feature for sizing the wing movable surfaces during the preliminary design of a novel vehicle. In fact, this work was performed in the framework of the FORMOSA and ATTILA projects, the former aimed at the design of the novel wing movable surface system of the NextGen Civil Tiltrotor aircraft (NGCTR), and the latter aimed at designing and testing a wind-tunnel model for the experimental verification of the whirl-flutter stability boundaries of the NGCTR. Both projects are developed within the framework of the Clean Sky 2-H2020 Program.

2. Description of the Coupled Software

2.1. Multibody Software MBDyn

MBDyn solves constrained dynamics of rigid and flexible systems by solving the Newton-Euler equations of motion of rigid and flexible bodies connected by kinematic constraints [8]. The problem is cast in first-order form as:

$$\mathbf{M}(\mathbf{x},t)\dot{\mathbf{x}} = \mathbf{p} \tag{1}$$

$$\dot{\mathbf{p}} = \boldsymbol{\phi}_{/\mathbf{x}}^T \lambda + \mathbf{f}_i(\dot{\mathbf{x}}, \mathbf{x}, t) + \mathbf{f}_e(\dot{\mathbf{x}}, \mathbf{x}, t) \tag{2}$$

$$\boldsymbol{\phi}(\mathbf{x}) = 0, \tag{3}$$

where $\mathbf{x}$ is the vector of the kinematic unknowns, $\mathbf{p}$ is the vector of the momentum unknowns, λ is the vector of the Lagrangian multipliers, $\mathbf{M}$ is a configuration (and possibly time) dependent inertia matrix, $\mathbf{f}_i, \mathbf{f}_e$ are arbitrary internal and external forces, $\boldsymbol{\phi}(\mathbf{x})$ are the algebraic constraint equations (only holonomic constraints are shown for conciseness), and $\boldsymbol{\phi}_{/\mathbf{x}}^T$ is the Jacobian matrix of the holonomic constraints with respect to the kinematic unknowns. Each structural node instantiates six force and moment balance equations (2), while only nodes with associated inertia properties instantiate the equations that define the momenta (1). The details associated with the handling of finite rotations are not made explicit; the interested reader is referred to Reference [8]. Additional states, associated with scalar fields (for example hydraulic pressure, temperature, voltage) and, thus, the associated differential and/or algebraic balance equations, can be taken into account through specialized sets of nodes.

The nodes that describe the kinematics of the structural problem can be connected either by elastic/viscoelastic internal forces (namely lumped structural components [43], beams [44,45], shells [46], Component Mode Synthesis (CMS) elements [47]) expressed by f_i, with a variety of viscoelastic constitutive laws, or by the kinematic constraints of Equation (3). Simple aerodynamics can be modeled by built-in elements that exploit the 2D strip theory model by look-up tables of the aerodynamic coefficients and classical rotor inflow models based on the momentum theory.

Coupling with an external solver is performed through external force element. This type of element communicates with the external solver by passing the kinematics of the model and receiving the corresponding loads. The coupling can be `loose` (i.e., explicit scheme) or `tight` (i.e., implicit scheme). For further details on the possible kind of coupling with an external solver, the reader is referred to Reference [48].

The external forces can be formulated directly in the absolute frame, or referred to a reference node. In the former case, operations are straightforward; in the latter, the kinematics are first reported in the reference frame of the reference node and then sent to the peer along with the motion of the reference node. The latter returns nodal forces and moments oriented according to the reference frame of the reference node.

2.2. Mid-Fidelity Aerodynamic Software DUST

DUST is an open source software designed by using object-oriented paradigms of the latest FORTRAN standards. DUST is a mid-fidelity aerodynamic software released under open source MIT license. The code was implemented using object-oriented paradigms of FORTRAN. The aerodynamic problem, based on incompressible flow hypothesis, is formulated by using the Helmholtz's decomposition of the velocity field that is given by the irrotational ($\mathbf{u}_\varphi$) and solenoidal $\mathbf{u}_\psi$ contributions of velocity as $\mathbf{u} = \mathbf{u}_\varphi + \mathbf{u}_\psi$. Different classical potential-based elements are implemented in the code, such as lifting line elements [49,50], typically used to model aerodynamics of slender lifting bodies as blades, surface panels [51], and vortex lattice elements (Reference [52], Section 10.4.3), typically used for aerodynamic modeling, respectively, of thick and thin solid bodies. Even if incompressible potential flow is considered in the code assumptions, compressibility effects are considered in the computations. In particular, a Prandtl-Glauert correction [53] is considered for steady aerodynamic loads calculation with surface panels and vortex lattice elements. On the other hand, lifting line elements intrinsically include compressibility and viscous effects introduced by Mach-depending tabulated sectional aerodynamic coefficients, obtained by experiments or by 2D RANS numerical simulations.

DUST simulations only require surface meshes, as the core of the mid-fidelity aerodynamic code is the vortex particles method (VPM) [31,32], a Lagrangian grid-free approach that is used to model the free vorticity of wakes and that does not require a volume mesh of the flow surrounding the object of investigation. Vortex particles reproduction of the flow enabled to obtain a robust representation of interacting wakes issued by lifting surfaces and bodies, as typically occurs in rotary-wing vehicles applications. In particular, in order to reduce the computational cost related to reproduce vortex particle interactions, a Cartesian Fast Multipole Method (FMM) [54], providing a background hierarchical decomposition of the domain into clusters of cells is implemented.

The code calculates aerodynamic loads using a different method according to the element considered to model the problem. For lifting line elements, the aerodynamic loads are calculated starting from the tabulated sectional aerodynamic coefficients used to solve their intensity. For vortex lattice elements, aerodynamic loads are calculated using the unsteady formulation of the Kutta-Joukowsky theorem. For surface panel elements, aerodynamic loads are calculated using the unsteady formulation of the Bernoulli theorem, considering the vorticity of the flow.

For a detailed insight on the mathematical formulation of the DUST code, including the implemented governing equations, the reader is referred to a recent work published by some of the present article authors [36].

2.3. Description of MBDyn-DUST Coupling

Communications between DUST and MBDyn are managed by preCICE (Precise Code Interaction Coupling Environment), a coupling library for partitioned multi-physics simulations, originally developed for fluid-structure interaction and conjugate heat transfer simulations. preCICE offers methods for transient equations coupling, communication means, and data mapping schemes. It is written in C++ and offers additional bindings for C, Fortran, MATLAB, and Python. preCICE (https://github.com/precice/) is a free software released under the LGPL3 license. While MBDyn uses its own C++, C, Fortran, and Python Application Programming Interface (API) to communicate with external software without any further modification to the C++ source code, no API was already available in DUST. Thus, new Fortran modules collecting all the classes, subroutines, and functions required by the adapter for preCICE were implemented to support coupling with DUST. The optional coupling with external codes was managed through pre-processor directives.

A new adapter was implemented for supporting the communication of all the kinematic variables (position, orientation, velocity, and angular velocity) plus forces and moments acting on the nodes of a MBDyn model exposed through an external structural force element. Figure 1 shows a scheme of the communication and information exchange, managed through the adapters for the two solvers.

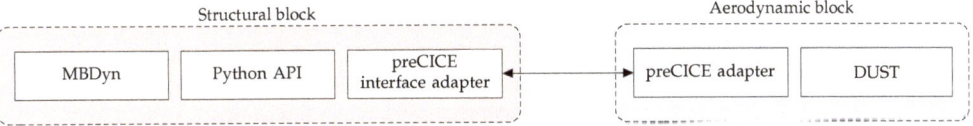

Figure 1. Scheme of the communication managed through adapters for MBDyn and DUST.

The interface between structural and aerodynamic grids is obtained as a weighted average of the distance between the nodes of the two grids and it is used for motion interpolation and consistent force and moment reduction. Figure 2 shows the q nodes of the structural grid, namely Q_q. The centers and the vertices of each aerodynamic mesh are, respectively, P_e and P_p, where e and p are the corresponding indices.

The kinematic variables, ϕ_p, of a point p positioned on the aerodynamic surface of a DUST component is evaluated as the following weighted-average,

$$\phi_p = \sum_q w_{pq} \phi_q, \qquad (4)$$

where ϕ_q is the same kinematic variable associated with the qth structural node of the MBDyn model.

Weights w_{pq} could be any set of non-negative real numbers, satisfying the normalization conditions

$$\sum_q w_{pq} = 1 \qquad \forall p, \qquad (5)$$

since they define the weighted average of the variables associated with the structural nodes q on the aerodynamic nodes p. These coefficients could be proportional to some negative power, defined as a user input, of a norm of the vectors $(P_p - Q_q)$. As an example, using the local coordinates in the reference configuration r_{pq}, the norm of these vectors can be defined as:

$$\|(P_p - Q_q)\|^2 := r_{pq}^T \, \mathbf{W} \, r_{pq}, \tag{6}$$

where $\mathbf{W}$ is a positive (semi-)definite matrix, providing an anisotropy degree of freedom to the user in defining the (semi-)norm [55]. Threshold values and maximum number of influencing weights are two criteria, defined as user inputs, used to restrict the average only to the significant structural nodes for each aerodynamic point.

2.3.1. Kinematic and Load Variables

The position of a point P in the global reference frame g of the aerodynamic surface is evaluated as

$$(P_p - O)^g = \sum_{Q \in I_p} w_{pq} \left\{ (Q_q - O)^g + \mathbf{R}_Q^{r \to g}(P_p - Q_q) \right\}, \tag{7}$$

where $Q \in I_p$ indicates the subset of structural points Q_q that belong to the I_p aerodynamic component, $(Q_q - O)^g$ is the distance from the origin of the Q_q structural point, and $\mathbf{R}_Q^{r \to g}(P_p - Q_q)$ rotates in the global coordinates the distance between the aerodynamic point and the structural one. Its angular velocity ω_P and velocity v_P, respectively, are

$$\omega_P = \sum_{Q \in I_p} w_{pq} \, \omega_Q, \quad v_P = \sum_{Q \in I_p} w_{pq} \{ v_Q + \omega_Q \times (P - Q) \}. \tag{8}$$

The aerodynamic forces and moments are evaluated at the evaluation points P_e located at the centers of each panel and then transferred to the structural nodes using the summation of forces and transport of moments as follows:

$$f_Q = \sum_{e \in J_Q} w_{qe} \, f_e, \quad m_Q = \sum_{e \in J_Q} w_{qe} \{ m_e + (P_e - Q_q) \times f_e \}, \tag{9}$$

where $e \in J_Q$ indicates the subset of evaluation points that belong to each sub-component J_Q, and the weights w_{qe} are calculated using Equations (6) and (7), by computing the distance between each structural node and evaluation point.

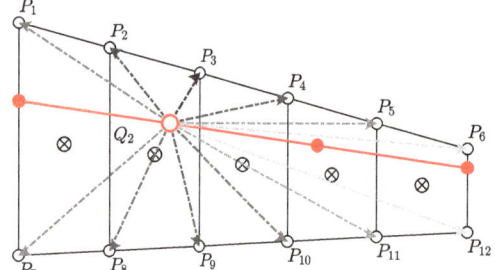 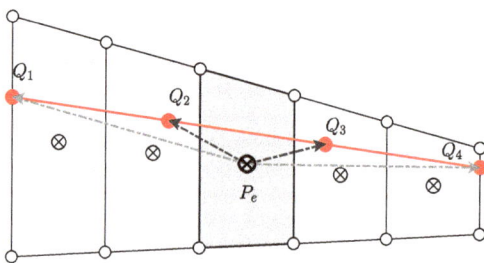

Figure 2. Scheme for motion interpolation (**left**) and force and moment transfer (**right**). Structural points are represented by red dots. Nodes of the aerodynamic mesh and panels centers are represented with plain dot and crosses, respectively.

2.3.2. Implementation of Software Coupling

Figure 3 shows the flow of information during the coupled simulation between the two solvers for an implicit tight serial scheme. First, the object `precice` of class `t_precice` is declared for handling a coupled simulation through the preCICE library. This object is used both for managing data communication, and for updating coupled components of the aerodynamic model. Then, the instance of DUST participating in the coupled simulation is created, reading the XML preCICE configuration file. After some preliminary operations, the mesh used to couple the codes is defined, and the fields involved in the communication are initialized. Initialization of the Fast Multiple kernels, wake, and linear system follows.

Before the time loop starts, communication is first established between the coupled codes. The time loop starts with the update of DUST's explicit aerodynamic elements (lifting lines and actuator disks). A checkpoint of the exchanged fields is stored, to be reloaded during sub-iterations of preCICE's implicit coupling. DUST then receives the kinematic variables of the structural nodes from the external software, MBDyn, and updates the surfaces of the coupled components and the near-field wake elements. The linear system is then updated and solved, calculating the strengths of the vortexes of the surface panels and vortex lattice elements.

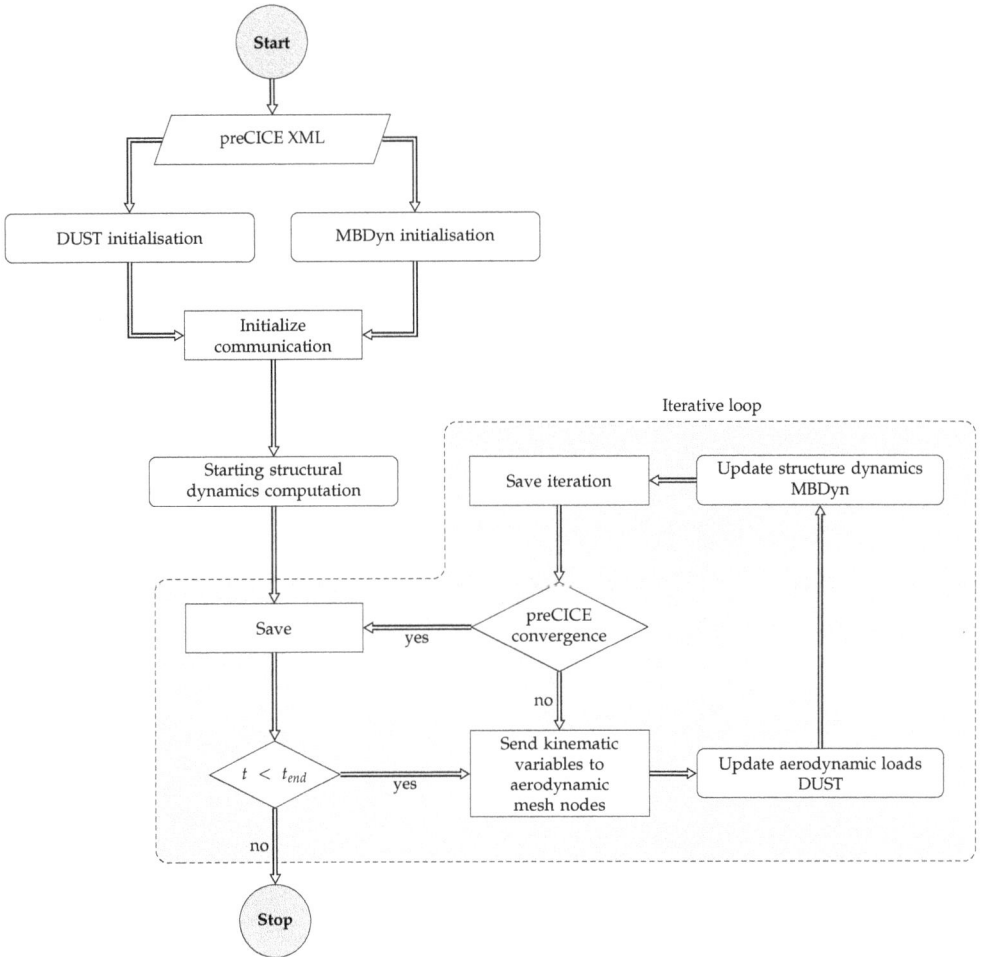

Figure 3. Flowchart of the implicit communication managed by preCICE between DUST and MBDyn [39].

The solution of the non-linear lifting line problem follows. The circulation Γ is computed by using analytical expression from the Kutta-Joukowski theorem, where lift coefficient is evaluated from tabulated sectional data. Once the intensity of the surface singularities has been evaluated, surface pressure distribution and elementary forces and moments are retrieved using the unsteady Kutta-Joukowski theorem for the vortex lattice and lifting lines elements, as well as the unsteady Bernoulli theorem for the 3D-panels.

Aerodynamic forces and moments are reduced to the nodes of the interface between the aerodynamic and structural meshes and sent to MBDyn. A convergence check on the kinematics variables follows. If convergence is not reached, the checkpoint fields are reloaded, and a new sub-iteration begins. If convergence is attained, the time step is finalized, saving the status and updating the wake and the geometry of the uncoupled components for the next time step.

2.4. Hinged Surfaces Modeling

Modeling the deflection of a control surface is essential for the simulation of aircraft maneuvers. In MBDyn, the deflection of a control surface can be easily modeled as a rigid/deformable body properly constrained to the fixed part of the vehicle to allow only its relative rotation about the hinge axis. In DUST, the possibility to include a control surface in the aerodynamic mesh has been only recently introduced. In the following, the description of the implemented model for hinged surfaces in DUST is introduced with a two-dimensional example first, and then extended to three-dimensional deformable components.

As outlined by the scheme in Figure 4 (left), in a two-dimensional problem the control surface can be defined in the local reference frame of the component, by means of the hinge axis position H, the chord-wise direction $\hat{\xi}$, and a blending region $[-u, u]$ introduced to avoid irregularities in the mesh as the surface rotates by an angle θ. Moreover, in the 2D modeling, the rotation axis $\hat{h}$ is assumed to be orthogonal to the plane of the airfoil.

As shown in Figure 4, a orthonormal reference frame for the hinge is defined with origin in H and axes $\hat{\xi}$, $\hat{\eta} = \hat{h} \times \hat{\xi}$. The position of a point with respect to this reference frame is

$$r = \xi \hat{\xi} + \eta \hat{\eta} + h \hat{h}. \tag{10}$$

Three regions are defined using the coordinates based on this reference frame:

1. $\xi \leq -u$: no influence of the control surface rotation;
2. $\xi \geq u$: rigid rotation about the hinge:

$$\Delta r = \sin \theta \hat{h} \times r + (1 - \cos \theta) \hat{h} \times \hat{h} \times r; \tag{11}$$

3. $-u \leq \xi \leq u$: blending region to avoid irregularities, defined as an arc of a circle whose center is located at point C and whose radius is:

$$\overline{CC'} = \frac{\overline{C'H}}{\tan \frac{\theta}{2}} \quad \text{where} \quad \overline{C'H} = u. \tag{12}$$

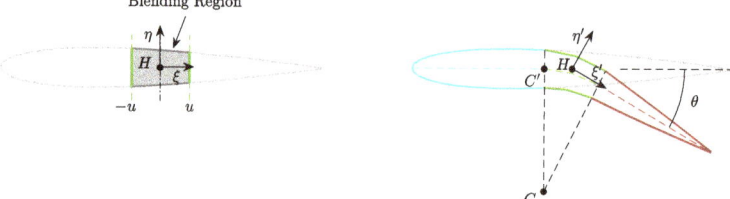

Figure 4. Scheme of the two-dimensional hinged surface configuration.

In a three-dimensional problem, the reference configuration of a control surface for a generic swept wing is defined in the wind axis reference frame of the component, as shown in Figure 5.

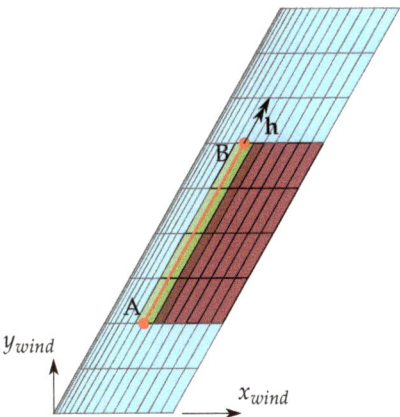

Figure 5. Hinge reference system for a swept wing.

The aerodynamic sections that are involved in the deflection of the control surface are, thus, the ones that satisfy the condition $y(A) < y(P) < y(B)$, where $y(P)$ is the ordinate of the $P_i - th$ aerodynamic mesh point expressed in the wind reference system. As in the 2D case, one can define the three regions for each stripe identified at the previous point. The y coordinate of the origin of the sectional reference frame is determined by linear interpolation between points A and B.

When the movable surface is coupled with a structural component in MBDyn, the orientation of the hinge in DUST nodes comes from the orientation of the nodes of the MBDyn model. The rotation axis is defined as $\hat{\mathbf{h}} = \dfrac{(B - A)}{\|B - A\|}$. Each point of the movable surface is linked to the hinge nodes, obtaining its kinematic variables as a weighted average of the motion induced by the rotation of the hinge nodes. Weights w_{ph} are evaluated using only the h components of the vectors connecting the control surface points to the hinge nodes.

The weights w_{ph} are then combined to the interpolations weight w described in Section 2.3 to allow the deformation of the structure, as shown in Figure 6. The weights are calculated for each point of the aerodynamic mesh, whereas the weights w_{ph} are specifically used to impose the rigid deflection of the movable surface.

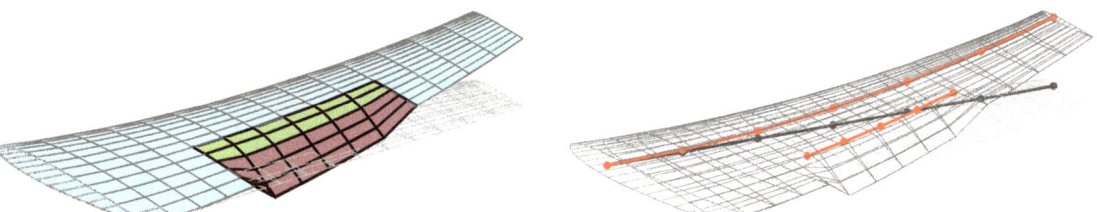

Figure 6. Control surface deflection for a deformable wing.

3. Validation of the Coupled MBDyn-DUST Software

3.1. Goland's Wing

This section presents the application of the coupled software to Goland's wing [56], which is widely used in the literature to test and validate aeroelastic codes. This low aspect ratio wedged wing ($AR \approx 3.33$) is also interesting to highlight the impact on flutter calculations of 2D and 3D aerodynamic models. Figure 7 shows the layout of the problem. EA indicates the elastic axis; CG indicates the center of gravity axis. U_∞ is the free-stream velocity. All the relevant geometrical and structural properties are reported in Table 1. They have been obtained from Reference [57].

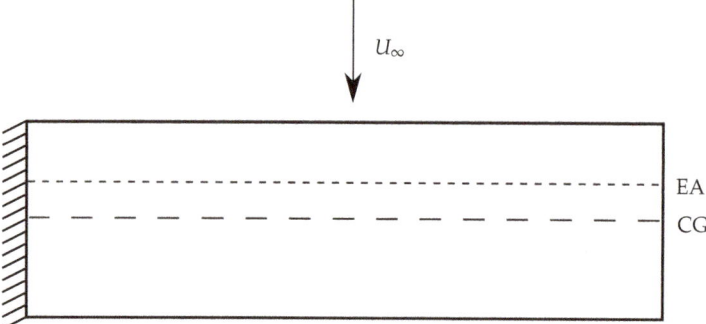

Figure 7. Goland's wing model layout.

Table 1. Goland's wing properties and flight condition [57].

Wing properties			
Half span	6.096 m	Inertia/unit span	8.64 kg m
Chord, c	1.8288 m	Mass/unit span	35.71 kg m^{-1}
Elastic axis	0.33·c	Torsional stiffness	0.99×10^6 N m^2
Center of gravity	0.43·c	Bending stiffness	9.77×10^6 N m^2
Flight condition			
Air density	1.020 kg m^{-3}	AoA Perturbation	0.05 deg

Two aerodynamic meshes were considered. In the first case, the wing was modeled as a flat plate using vortex lattice (VL) elements, while, in the second case, it was modeled with surface panels (SP) reproducing its geometrical shape and thickness. The results of a convergence analysis on the computed aerodynamic loads indicated the need to use 30 elements in the span-wise direction to obtain a correct spatial discretization for both models. The VL flat mesh requires 30 uniform divisions in the chord-wise direction, while the SP model requires 30 divisions for the lower and upper side of the wing, with half-cosine refined distribution at leading edge. The structural model built in MBDyn consists of four beams based on a C^0 beam discretization based on the finite volume concept proposed in Reference [44]. A satisfactory number of beam elements was obtained by satisfying a convergence requirement on the first four modes of the wing, as indicated by comparison of the frequencies of the first four normal vibration modes of the wing computed by MBDyn with respect to Goland's work [56] and NASTRAN shown in Table 2.

Table 2. Comparison of the first four natural frequencies computed for the Goland's wing [56].

	Goland [56] Hz	NASTRAN Hz	MBDyn Hz
1st Bending	7.66	7.66	7.66
1st Torsion	15.24	15.24	15.21
2nd Bending	38.80	38.59	38.54
2nd Torsion	55.33	54.84	54.79

In the present study, tight coupled time-marching simulations were performed using a time discretization of 0.001 s and considering for the evolution of the wake particles a fourteen chords long box behind the wing, resulting in a developed wake made by about three thousand vortex particles. The computation time to perform a time-marched coupled simulation with a total duration of one second, using a workstation equipped with an Intel®

Core™ i9-9980XE processor running on a base frequency of 3.00 GHz, with 18 physical cores and 2 threads for each core, was about 19 min for VL and about 29 min for SP.

To study the flutter instability, a non-zero angle of attack of 0.05 deg was introduced as perturbation, as in Reference [58]. The frequency and damping of the response were identified from the time history of the wing-tip deflection using the matrix pencil estimation (MPE) method [59]. Figure 8 presents the z-displacement of the last structural external node across the flutter onset computed with the SP model. The red line corresponds to a stable damped response, whereas the blue line shows the condition of incipient flutter, in which a constant amplitude free oscillation is reached. The green line shows the unstable response at a speed greater than the flutter speed. A corresponding representation of the deformed mesh associated with the bending-torsion instability and the related distribution of the pressure coefficient is shown in Figure 8 (left). These results indicate the ability of the coupled simulation to correctly capture fixed-wing flutter.

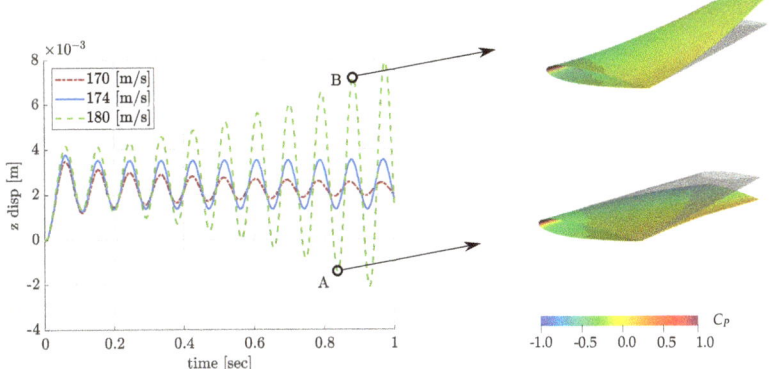

Figure 8. Time history of the Goland's wing-tip deflection evaluated with surface panels aerodynamic mesh at three different wind speeds.

Table 3 reports a comparison of flutter speed and frequency computed by several independent authors. Results computed with the coupled code are in quite good agreement with those obtained by similar codes using 3D aerodynamic models [57,58,60]. In detail, the discrepancy with the results obtained with the same MBDyn structural model, but using its built-in aerodynamics based on two-dimensional unsteady strip theory, indicates the superior capability of the coupled code in the investigation of aeroelastic problems. The comparison of the results from the literature reported in Table 3, obtained with 2D and 3D models, confirms the need of a three-dimensional aerodynamic model for a correct and realistic flutter analysis of low aspect ratio wings.

Table 3. Comparison of flutter speed and frequency computed for Goland's wing.

Author	Model	V_f ms^{-1}	f_f Hz
Goland [56]	Analytical	137.2	11.25
Patil et al. [57]	Intrinsic beam + strip theory	135.6	11.17
Wang et al. [60]	ZAERO	174.3	-
Wang et al. [60]	Intrinsic beam + UVLM	163.8	-
Murua et al. [58]	SHARP, Displacement beam + UVLM	165	10.98
Present Work	MBDyn's built-in strip theory	135.1	11.07
Present Work	DUST (VL)-MBDyn	168.2	10.84
Present Work	DUST (SP)-MBDyn	174.2	11.06

To evaluate the capabilities of the coupled code using the different employed aerodynamic models, Figure 9 presents the frequency and damping of the first beam torsional mode of the wing as functions of the free-stream velocity.

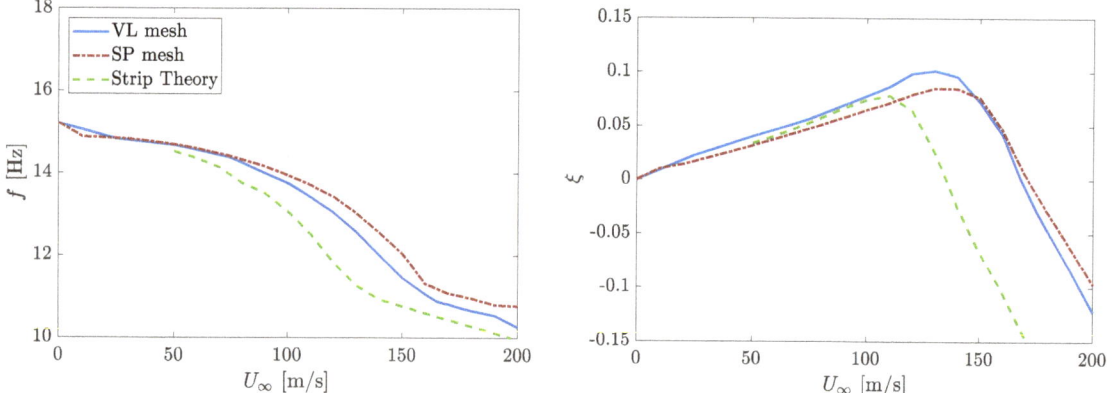

Figure 9. Frequency and damping versus velocity for Goland's wing. Coupled simulations results (VL and SP mesh) and MBDyn results with 2D Strip Theory aerodynamics.

The numerical results of the coupled simulations using a panel mesh (SP) show a slightly higher aerodynamic damping with respect to those obtained using VL. A predicted flutter speed increase of approximately 3.7% is observed (Figure 9 right). Nevertheless, the quite limited differences in the results of the two models indicate that, for simple configurations, without complex aerodynamic interaction between bodies, a vortex lattice mesh is more convenient than surface panels, as the computational cost reduces significantly.

3.2. XV-15 Proprotor in Hover

A rotary-wing test case was also considered to validate the MBDyn-DUST coupling. It consists of the XV-15 proprotor equipped with metal blades, a three-blade stiff-in-plane rotor with gimballed hub. For the work presented here, the gimbal universal joint has been modeled as ideal homokinetic joint, neglecting the 2/rev components caused by rotor flapping [61]. This test was selected thanks to the availability in the literature of the rotor geometrical and structural data [62,63] and of experimental data in hover condition. The experimental data were obtained during the test campaign that took place at the Outdoor Aerodynamic Research Facility (OARF), which is described in the work by Felker et al. [64].

The layout of a control chain representative of that of the XV-15 proprotor is shown in Figure 10. It was modeled using information from Reference [63]. The flowchart in Figure 11 details the blade pitch control system implemented in MBDyn. The role of each component is:

- `Pylon`: this node represents the actual connection between the pylon extremity and the rotor; when the isolated rotor is analyzed, this node is grounded.
- `Control`: through this node the controls (cyclics and collective) are prescribed. To decouple longitudinal and lateral cyclic inputs, the node is defined in a reference system that is rotated about the shaft by the angle $\psi_{sp} = \tan^{-1}(x_{sp}/y_{sp})$, where x_{sp} and y_{sp} are the in-plane components of the point where the pitch link is connected to the swashplate.
- `Fixed Swashplate`: this node represents the non-rotating portion of the swashplate. Its in-plane displacements and rotation about the shaft axis are constrained to the `Control` node. To account for the flexibility of the control chain, it is connected to the `Control` node by three equally radially spaced elastic rods.

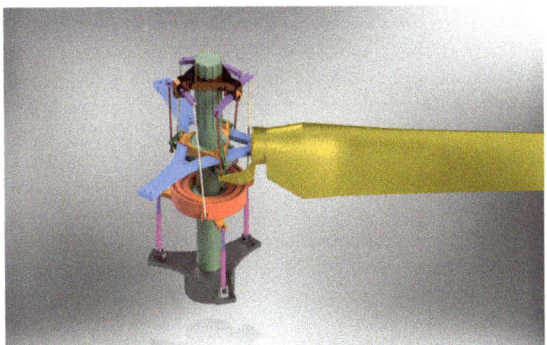

Figure 10. Layout of the XV-15 proprotor control chain.

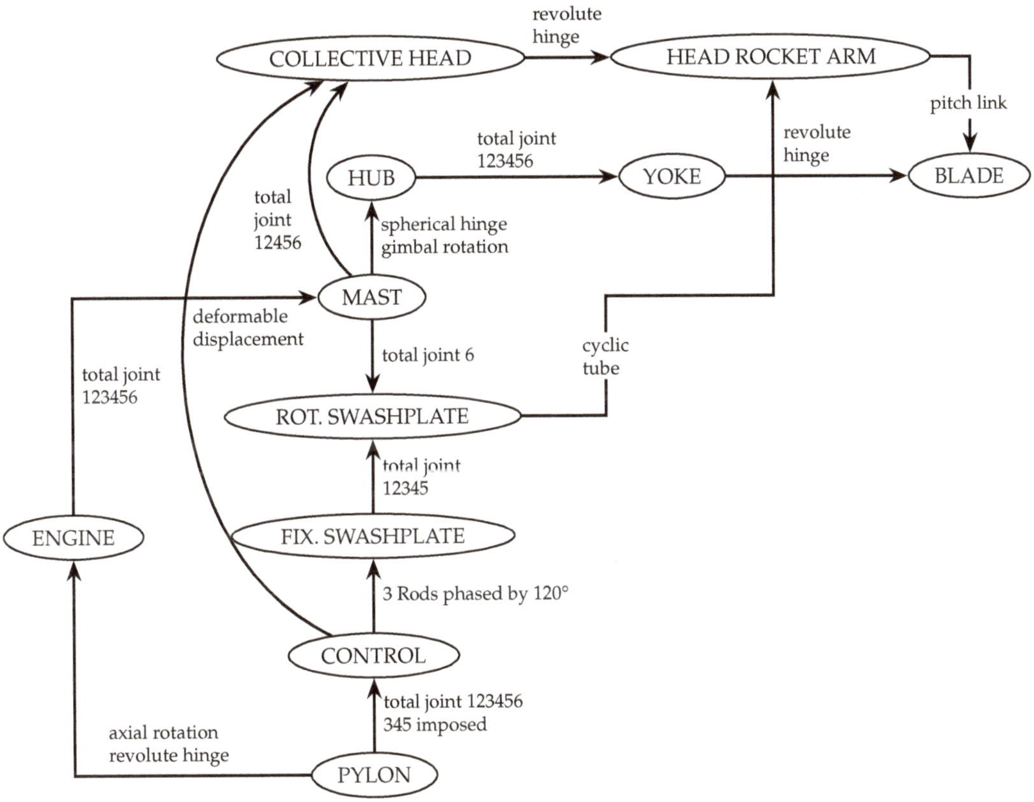

Figure 11. Flowchart of the MBDyn model of the XV-15 proprotor, particularly showing the individual blade pitch control system components and their connections for the dual control path.

- `Rotating Swashplate`: this node is connected to the `Fixed Swashplate` by a revolute hinge, and to the `Mast` node (defined later) by a joint that makes it rotate along with the shaft itself.
- `Collective Head`: this node is connected to the `Control` node by a deformable spring along the shaft axis, to account for the flexibility of the collective control path; it is positioned in the rotating reference system.

- Head Rocket Arm: this node is connected to the Collective Head node by a revolute hinge, and to the rotating swashplate by means of a rod (cyclic tube). Then, through the pitch link, the cyclic and collective commands are transmitted to the blade.
- Engine: this node represents the torsional inertia associated with the engine.
- Mast: this node drives the Hub (defined later) and the Rotating Swashplate nodes. It is connected to the Pylon node by means of a revolute hinge, whose angular velocity is prescribed.
- Hub: this node is constrained to the Mast node by a spherical hinge and a gimbal joint [65]; the combination of these two joints represents an ideal constant velocity joint.
- Yoke: this component represents the rigid body that connects the hub with the blade. The blade-to-yoke connection is obtained using a single load path through a revolute hinge.
- Blade: this node condenses both the inertial and aerodynamic properties of each proprotor blade.

All rotor data were taken from the original CAMRAD II model presented in Reference [62], here reported for clarity in Table 4. Table 5 presents the blade span-wise airfoil distribution, from Reference [64].

Table 4. XV-15 main rotor data [62].

Rotor Data		
Blades	3	
Solidity	0.0891	
Radius	3.81	m
Precone	2.5	deg
Chord	0.3556	m
Nominal speed	589	RPM

Table 5. XV-15 Blade airfoil distribution [64].

Airfoil Data		
Profile	start	end
NACA 64-935	0.09	0.13
NACA 64-528	0.13	0.34
NACA 64-118	0.34	0.655
NACA 64-(1.5)12	0.655	0.9
NACA 64-208	0.9	1

For validation purposes, simulations using MBDyn alone, DUST alone, and the coupled solvers have been performed. In detail, in the MBDyn alone simulations the blade aerodynamics was modeled with the Blade Element/Momentum Theory using a Glauert model [66] for the induced velocity. A tip loss correction was introduced to scale the value of the normal aerodynamic force with a prescribed span-wise weight.

The aerodynamics of each blade was modeled in DUST using lifting line elements, naturally encompassing both compressibility and viscous effects, as described in detail in Reference [36]. This aerodynamic model provides accurate results on high aspect ratio bodies, as blades are, while being computationally very efficient, as shown in Reference [67].

The DUST aerodynamic model of the full-scale XV-15 proprotor was built considering the airfoil geometry, chord, twist, sweep, and dihedral distributions reported in Reference [68]. Lifting line elements were used for the aerodynamic model of the three blades. The tabulated aerodynamic performance coefficients of the blade airfoils were calculated using XFOIL [69] in the range of angle of attack before stall. The two-dimensional aerodynamic load curves are extended to the $[-180° : 180°]$ range of angles of attack using the methods described in References [2,70]. While the MBDyn model applies a non-smooth transition between adjacent airfoil sections, DUST interpolates the aerodynamic properties of the airfoil sections used to build the blade model. A convergence analysis on computed rotor thrust and torque indicated the need to use 40 lifting line elements for the spatial discretization of each blade and a time discretization of 100 time steps to simulate a complete rotor revolution. For the evolution of the wake particles, a 5 rotor radius long box was considered in the simulations, resulting in a developed wake of around 52 thousand vortex particles. The computational time required to perform a DUST alone simulation over 5 complete rotor revolutions was about 324 s, while tightly coupled simulations using the same aerodynamic DUST model took around 395 s using the 18-core workstation described in Section 3.1.

Figure 12a shows the solidity-scaled thrust coefficient C_T/σ versus torque coefficient C_Q/σ curves computed for the XV-15 rotor in hover with MBDyn alone, DUST alone, and the coupled MBDyn-DUST solver, compared with experimental data reported in Reference [64]. It is worth noticing that:

(i) the curves from the MBDyn-DUST coupled simulation agree fairly well with experimental data, particularly at high values of thrust coefficient, thus indicating the suitability of the coupled tool to improve accuracy in the evaluation of rotor performance with a quite limited computational effort;

(ii) on the contrary, the curves from the MBDyn alone and DUST alone simulations both appreciably depart from experimental data and the coupled simulation results, although somehow capturing the observed trends;

(iii) the curves from the MBDyn alone and DUST alone simulations appear to be in good agreement with each other, especially at high values of thrust coefficient;

(iv) however, this agreement actually tells two different stories: MBDyn alone uses a poor aerodynamic model, but correctly describes the kinematics of the problem, whereas DUST alone uses a better aerodynamic model but with prescribed kinematics; it appears that, for this specific problem, a more accurate structural model and a more accurate aerodynamic model alone give a comparable contribution to model fidelity; clearly, this observation can hardly be generalized;

(v) it also appears that, in the MBDyn alone analysis, the presence of an inflow model and of standard, well understood and accepted empirical corrections, can substantially improve the quality of the prediction, compared to the significant error that was observed in Goland's wing flutter analysis.

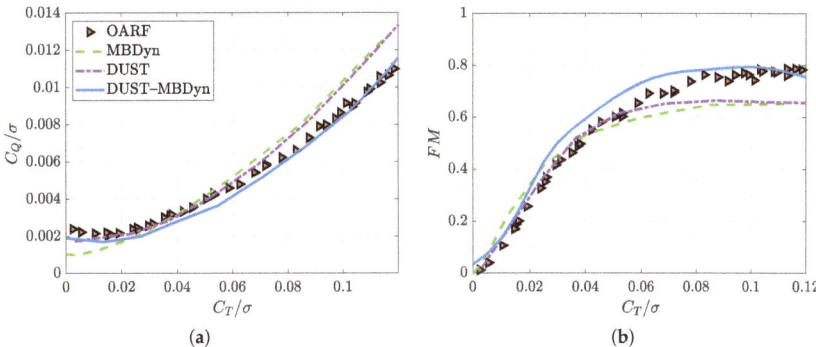

Figure 12. Comparison of results obtained for the XV-15 full-scale single proprotor in hover: (**a**) torque coefficient C_Q/σ versus thrust coefficient C_T/σ; (**b**) figure of merit (FM) versus thrust coefficient C_T/σ.

4. Simulation of the Roll Maneuver on the Complete XV-15 Tiltrotor

A full-span aeroelastic model reproducing the Bell XV-15 research aircraft equipped with rigid metal blades [62] is considered for the coupled roll maneuver simulations. The purpose of this analysis is to show the ability of the code to model complex problems, particularly those involving the deflection of movable control surfaces, and to give an indication of the possible outcome of the proposed novel coupled numerical tool.

4.1. Numerical Model

The numerical model of the XV-15 tiltrotor is built considering the full scale dimensions and components of the aircraft. The model includes the fuselage, the horizontal and vertical tail-planes, the wing equipped with control surfaces (flap and flaperon) and the two proprotors with the corresponding nacelles. The main geometric characteristics of the entire vehicle, including the airfoils, are reported in Table 6.

Table 6. Geometric characteristics of the XV-15 [41].

	Wing	Horizontal Tail	Vertical Tail
Airfoil	NACA 64A223	NACA 64015	NACA 0009
Span	9.8 m	3.91 m	2.34 m
Mean aerodynamic chord	1.60 m	1.20 m	1.13 m
Sweep ($c/4$)	$-6.5°$	$0°$	$31.6°$
Dihedral	$2.0°$	$0°$	-
Incidence	$3.0°$	$0°$	$0°$
	Flap	Flaperon	
Span along hinge line	1.30 m	2.40 m	
Chord/Wing chord	0.25	0.25	
Maximum deflection	$75°$	$47°$	

The aerodynamic model uses different types of aerodynamic elements: lifting lines for the blades of the proprotor, using the same spatial discretization of Section 3.2, and surface panels for all the other aerodynamic components: 2514 elements are used for the fuselage, 1620 for the tail planes, 835 for each nacelle, and 3500 for the wing. The aerodynamic mesh of the complete XV-15 tiltrotor is shown in Figure 13.

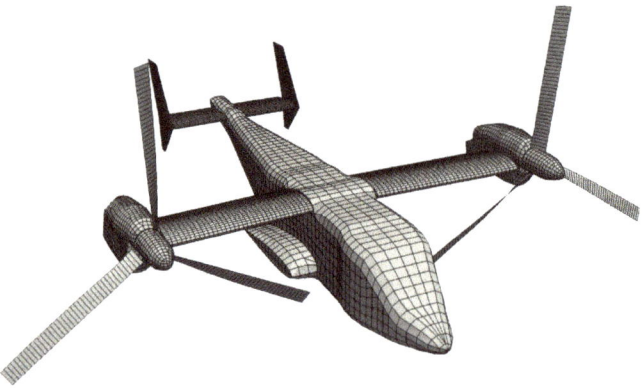

Figure 13. Aerodynamic mesh of the XV-15 tiltrotor model.

The dynamic model includes:
- the wing, modeled as a rigid body, including flaps and flaperons;
- the fuselage and the empennages, modeled as rigid bodies, including the rudders and the elevator control surfaces;
- the pylon/nacelle system, mounted to the wing-tip; it can be tilted with respect to the wing to reproduce tiltrotor in airplane mode (APMODE), helicopter mode (HEMODE), or in any intermediate attitude;
- the rotor with the exact kinematics of the blade pitching mechanism, as described in Section 3.2.

The inertial structural data used for the multibody model were taken from the CAMRAD II and NASTRAN models presented in Reference [62]. Table 7 reports the inertia data of the complete aircraft considering a reference system having the x-axis pointing towards the tail, the y-axis aligned with the right wing, and the z-axis pointing upward according to the right-hand rule.

Table 7. Center of mass, mass, and inertia tensor of the complete aircraft [62].

x_{CG}	y_{CG}	z_{CG}	M
7.214 m	0 m	−0.425 m	5896.8 kg

I_{xx}	I_{yy}	I_{zz}	I_{xy}	I_{xz}	I_{yz}
67,512.6 kgm^2	335,546 kgm^2	398,235 kgm^2	4.4 kgm^2	23,136.5 kgm^2	−0.5 kgm^2

From the structural point of view, the flaperon is modeled as a rigid body connected to two nodes representing the hinges. The first node is constrained through a spherical hinge, while the second is constrained by an inline joint. The rotation of the flaperon is obtained by imposing a prescribed rotation on the second constraint. The combination of these three joints makes the constraint statically determined. The line connecting the two nodes identifies the axis of rotation of the movable surface; the aerodynamic mesh is deformed according to the methodology illustrated in Section 2.4.

To evaluate the aerodynamic effects of the tail-planes and their interaction with the wake of the rotors on roll performance, coupled simulations were performed for three different aircraft configurations, according to Table 8 and Figure 14. First, the airframe only was considered; then, the tail-planes were added, and, finally, the rotors, increasing the complexity and the completeness of the model. The structural model and the relative mass properties were the same for the three simulated configurations.

Table 8. Aircraft configurations tested for roll maneuver coupled simulations.

Configuration	Airframe	Tail	Rotor
Configuration I (C.I)	✓	✗	✗
Configuration II (C.II)	✓	✓	✗
Configuration III (C.III)	✓	✓	✓

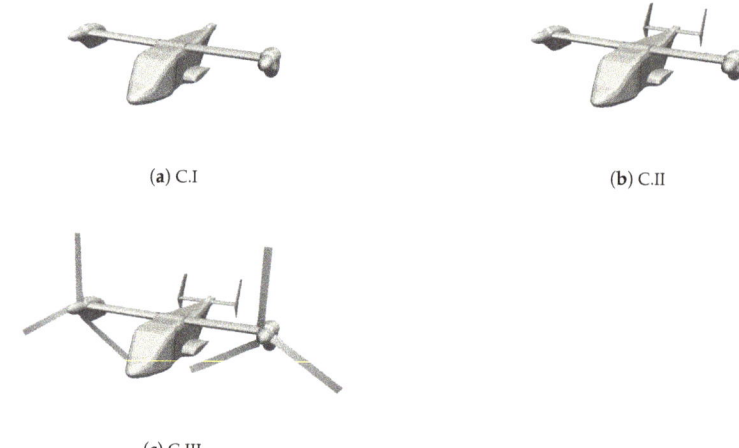

(a) C.I (b) C.II (c) C.III

Figure 14. Geometry of aircraft configurations tested for roll maneuver coupled simulations.

A trimmed flight condition reported by Ferguson [71] was considered for the roll maneuver coupled simulations. The flight condition and trim parameters are reported in Table 9.

During the roll maneuver simulations, the flaperons are deflected according to a step function from $\delta_a = 0°$ to $\delta_a = \pm 20°$. The control surfaces start moving after 0.5 s from the beginning of the simulation, to make sure the maneuver starts after any initial aerodynamic transient vanish. At the same time, the roll degree of freedom of the entire model is unlocked. In the simulations, the aircraft rolls about the longitudinal axis, positive starboard (right) wing up. Yaw rotation is about the vertical body axis, positive nose left, while pitch rotation is about the axis normal to longitudinal plane of symmetry, positive nose up.

Table 9. Flight condition parameters used for roll maneuver simulation [71].

Flight Condition		
Air density	1.225	$\mathrm{kg\,m^{-3}}$
Speed	72.022	$\mathrm{m\,s^{-1}}$
Trim parameters		
Mode	Airplane	
Pitch angle	6.944	deg
Rotor speed	517	RPM
Rotor collective	29.5015	deg
Elevator	−1.2398	deg

Considering a tightly coupled simulation using 145 time steps over a complete blade rotor revolution, the computational time to perform a one second long simulation for the full vehicle C.III was about 8 h using the previously mentioned 18-core workstation. In detail, the full developed wake for C.III consisted of approximately 84 thousand particles, considering a box domain twice the length of the aircraft in the x-direction. No convergence studies were performed for the full vehicle coupled simulation. Indeed, the choice of the selected spatial and time discretizations for the full vehicle was dictated by best practices (see Reference [37]) and by the will to limit the computational effort of the simulations.

Indeed, one of the goals of the present activity is to show the capabilities of the novel coupled numerical tool for the preliminary design of innovative rotary-wing aircraft configurations that requires a wide number of numerical simulations to reproduce the different attitudes of their flight mission.

4.2. Results and Discussion

The present section reports the main results of the coupled simulations for the complete vehicle, starting from the discussion of the response to the roll maneuver in terms of flight mechanics performance, here intended in terms of bank angle and roll rate for a prescribed aileron deflection.

Figure 15a shows the comparison of the bank angle ϕ evolution during the simulated roll maneuver for the three aircraft configurations tested. In particular, the figure clearly shows that the aerodynamic effects of the tailplanes and rotor wakes changes the slope of the bank angle curve and influences the roll maneuver performance.

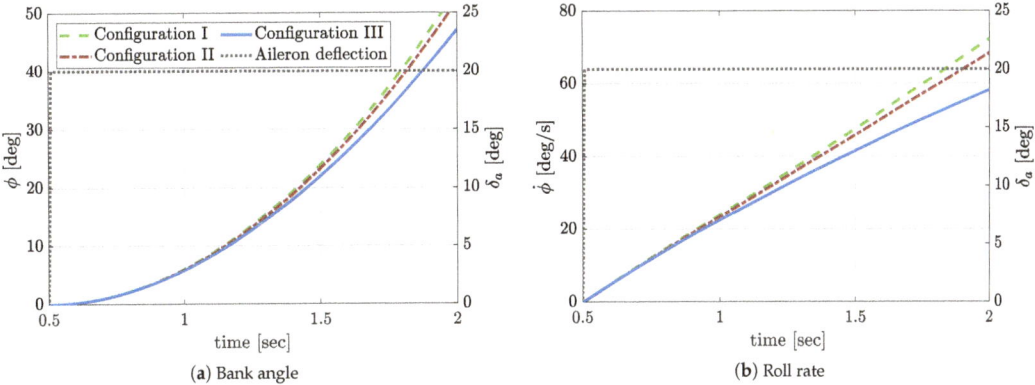

Figure 15. Comparison of: (**a**) the computed bank angle evolution and (**b**) roll rate evolution for the tested configurations.

Specifically, roll performance appears to degrade when additional contributions to the aerodynamics are considered. Table 10 shows the negative percentage differences between the time computed to bank by a 45° angle for aircraft C.I and C.II with respect to the complete aircraft equipped with rotors, i.e., C.III.

Table 10. Percent difference of time to bank 45° for aircraft C.I and C.II with respect to C.III.

Configuration	Δ%
C.I	−4.46 %
C.II	−3.33 %

The performance reduction due to the tail surfaces amounts to about 1%; it is explained by the change in angle of attack due to the roll rate, which produces a negative roll moment contribution. The performance reduction due to the proprotor aerodynamics, confirmed by the comparison of roll rate evolution ($\dot{\phi}$) presented in Figure 15b, is related to a backward tilting of the rotor induced by the component of reference velocity associated with roll rate in the rotor disk plane. It is also observed that the aircraft roll causes an opposite variation of the rotors thrust, as shown in Figure 16.

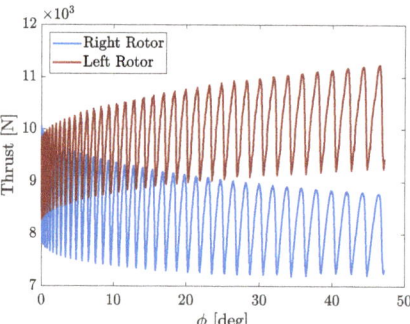

Figure 16. Rotor thrust with respect to bank angle during the roll maneuver.

This imbalance between the right and left rotor thrust generates a non-negligible contribution to the yaw moment reaction at the revolute joint that grounds the aircraft. Since the thrust of the left rotor increases and that of the right one decreases, the sign of the yaw moment perturbation is negative.

Figure 17a shows the measured yaw moment reaction at the ground joint. Since the position of this constraint is fixed during the simulation, the purpose of these results is not to establish the actual behavior of the aircraft but to estimate the impact of the different vehicle parts on this quantity.

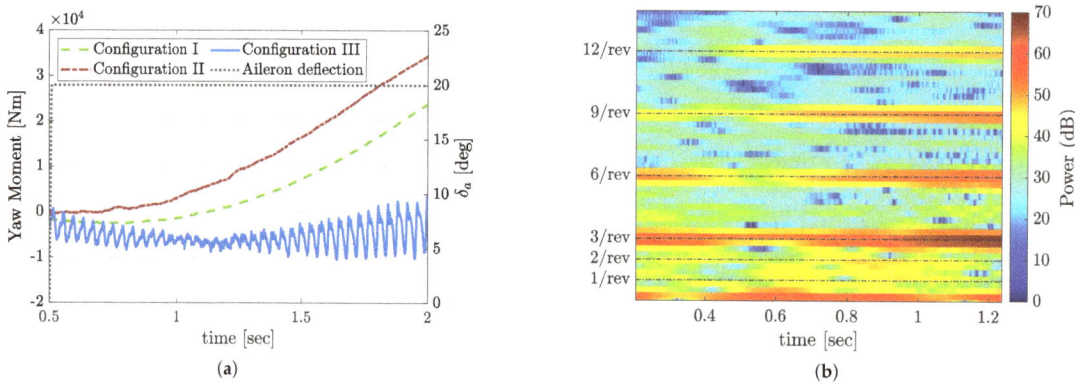

Figure 17. Yaw moment during the roll maneuver. (**a**) Yawing moment reaction evolution during roll. (**b**) Spectrogram of yaw moment time history for C.III.

The analysis shows that tailplanes generate a proverse yaw contribution compared to the simplest C.I. On the contrary, the introduction of the aerodynamic contribution of the proprotors generates an adverse yaw moment. This effect is related to the opposite variation of the rotor thrusts highlighted in Figure 16.

Considering dynamic oscillations of the aircraft during the roll maneuver, a quite complex response is observed for C.III, the full aircraft configuration with rotors. In particular, the spectrogram of the yaw moment time history computed for C.III shown in Figure 17b clearly identifies the correspondence of these oscillations with the multiples of the rotor N/rev. Furthermore, it is evident that almost all harmonics increase in power as the simulation advances over time. The development of these aerodynamic states is due to the capability of the aerodynamic solver DUST to capture aerodynamic interactions.

Thus, in order to deeply understand the effect of possible aerodynamic interactions on the maneuver, the effect of the two proprotors on the normal force acting on the wing is investigated. Figure 18 shows the convention adopted for the azimuthal angle of the blade ψ.

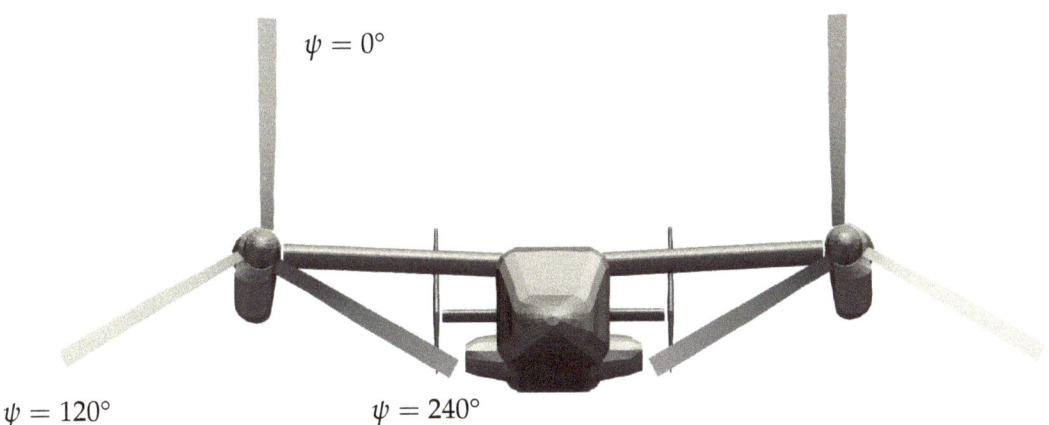

Figure 18. Convention adopted for the rotor blade azimuthal angle ψ.

Figure 19 shows the azimuthal histories of the normal force computed on the left and right side of the wing for the three investigated configurations over a complete rotor revolution at the same bank angle $\phi = 30°$. As one may expected, for aircraft configurations I and II, there are no oscillations due to the aerodynamic loads, as the proprotor's wake is absent. A slight difference between results obtained for configurations C.I and C.II can be noted on the right wing only, as the introduction of the tail provides an increase of normal aerodynamic load during the maneuver. The blue line curves, corresponding to the wing normal force for C.III, show the three peaks related to the passage of the blades in front of the wing. This comparison shows the capability of the mid-fidelity aerodynamic model to reproduce the detailed effects of proprotor wake interaction on the wing, providing an increase of the aerodynamic load on both sides. This effect is mainly due to the acceleration of the flow impinging the wing caused by the proprotor inflow, which provides a local increase of the wing loading.

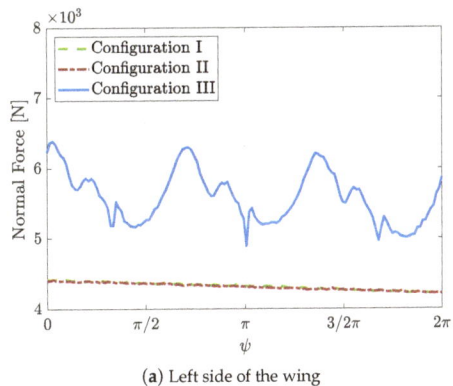
(a) Left side of the wing

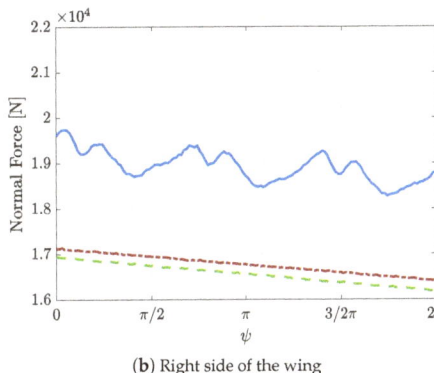
(b) Right side of the wing

Figure 19. Azimuthal history of wing normal force during a rotor revolution at bank angle $\phi = 30°$, (a) left side and (b) right side.

Figure 20 shows the span-wise distribution of the normal force acting on the wing for the three simulated aircraft configurations at bank angle $\phi = 30°$. Results for C.III are plotted at three different blade azimuth angles ψ. The span-wise wing load comparison confirms the increase of the normal force due to interaction of the proprotor wake. The effect on the aerodynamic loading due to flaperons deflection is apparent, as well. Indeed, the aerodynamic model accurately captures the reduction of wing loading on the left, due to the control surface negative deflection, and the increase of loading on the right due to the control surface positive deflection. Focusing on C.III simulation results, the span-wise load variation slightly changes with blade azimuth angle ψ, due to the interaction between the tip vortices released by the proprotor blades and the wing. Indeed, the normal force acting on the wing presents a peak at $\psi = 0°$, as also shown in the time history of Figure 19.

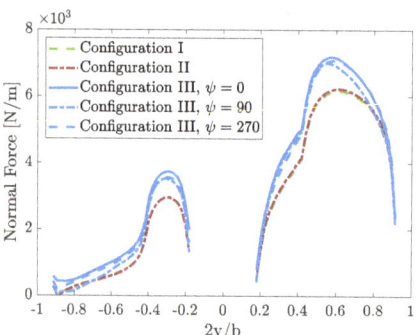

Figure 20. Span-wise normal force distribution on the wing at bank angle $\phi = 30°$.

A more detailed insight into the interactional effects of the proprotors on the wing can be deduced from Figure 21. The figure presents a visualization of the flow field behind the proprotors through Q-criterion iso-surfaces and the distribution of the pressure coefficient C_P over the aircraft surfaces for the three tested configurations. In detail, the flow field representation for C.III clearly shows the interference of the proprotor helical system with the wing. This interaction mainly provides an up-wash due to the blade tip vortex encountering the inboard region of the wing [37] and a consequent local increase of pressure with respect to C.I and C.II, as shown by the larger and more intense pressure region computed over the wing surface for the aircraft configuration equipped with proprotors (see Figure 21c).

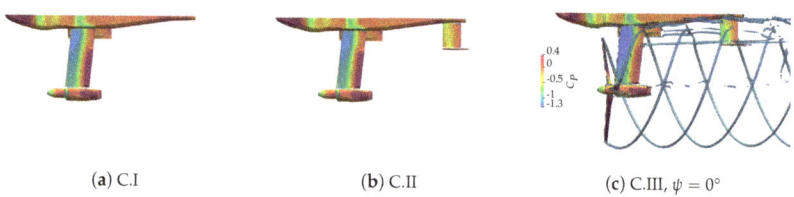

(a) C.I   (b) C.II   (c) C.III, $\psi = 0°$

Figure 21. Coupled simulation results for the three configurations at bank angle $\phi = 30°$: pressure coefficient C_P contours over the aircraft surface for the three configurations and iso-surface of Q-Criterion ($Q = 1200$) for flow field description for C.III.

Figure 22 compares the pressure coefficient C_P distribution evaluated at the mid-span section of the flaperon. The C_P curves show that interaction with the proprotor wake is responsible for an increase of the suction peak on the upper surface at the leading-edge region of the airfoil. Otherwise, the pressure coefficient distribution is quite similar for all configurations in the aft portion of the airfoil, which corresponds to the deflected flaperon.

A higher aerodynamic loading of the complete aircraft wing results in C.III with respect to C.I and C.II due to aerodynamic interaction with the proprotor's wake.

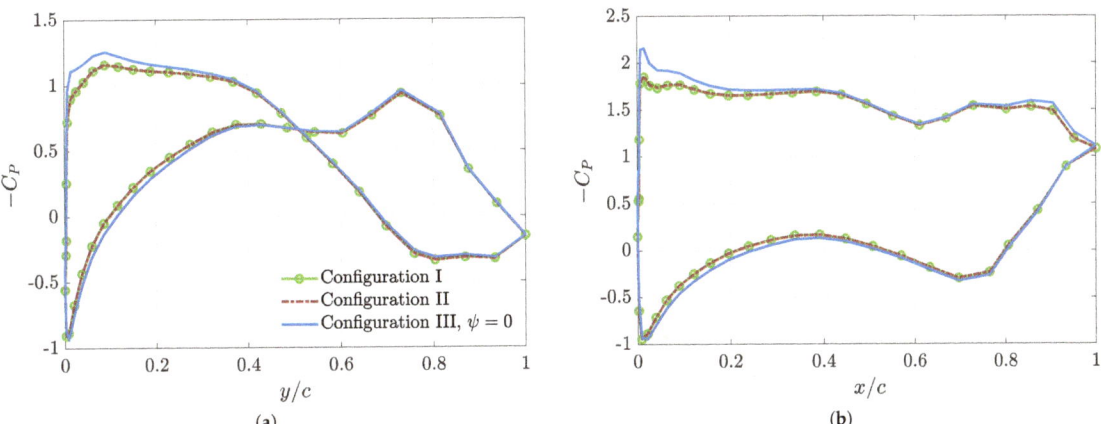

Figure 22. C_P distribution at flaperon mid-span for the three configurations at bank angle $\phi = 30°$. (**a**) C_P left wing; (**b**) C_P right wing.

Finally, the effect of the aerodynamic interference of the wing on the proprotors is investigated. Figure 23 shows the contours of the blade lift coefficient (C_L) computed over a complete rotor revolution for different bank angles ϕ. The polar plots show that the mid-fidelity aerodynamic model captures quite well the variation of the thrust force distribution in the region corresponding to the passage of the blade in front of the wing. Considering the blade lift coefficient distributions on the proprotors at the same bank angle, a quite similar distribution is observed between the two rotors, except for a small difference in the innermost area of the disk, which is related to the different position of the aileron (positive deflection on the right, negative deflection on the left); see Figure 23a,b referring to the moment in which the ailerons are deflected, and the bank angle is null. As the maneuver progresses and the bank angle increases, the lift distribution on the two proprotors presents more differences. In particular, focusing the attention on the right rotor, it is evident that increasing the bank angle between $\phi = 0°$ (see Figure 23a) and $\phi = 45°$ (see Figure 23g), the lift coefficient positive variation in the outer region of the blade decreases in the range of blade azimuthal angle between $\psi = 210°$ and $\psi = 270°$, characterized by the passage of the blade in front of the wing. The left rotor behaves in the opposite way showing, as the bank angle increase, a positive variation of the lift coefficient in the outer region of the blade for the same azimutal blade angle range (see Figure 23b–h). In detail, the magnitude of this lift variation within the maneuver is greater for the left proprotor. This opposite trend of the proprotors local loads reflects the physics of the unbalance of the two proprotors thrust discussed in Figure 16 and the consequent growth of an adverse contribution to the yaw moment.

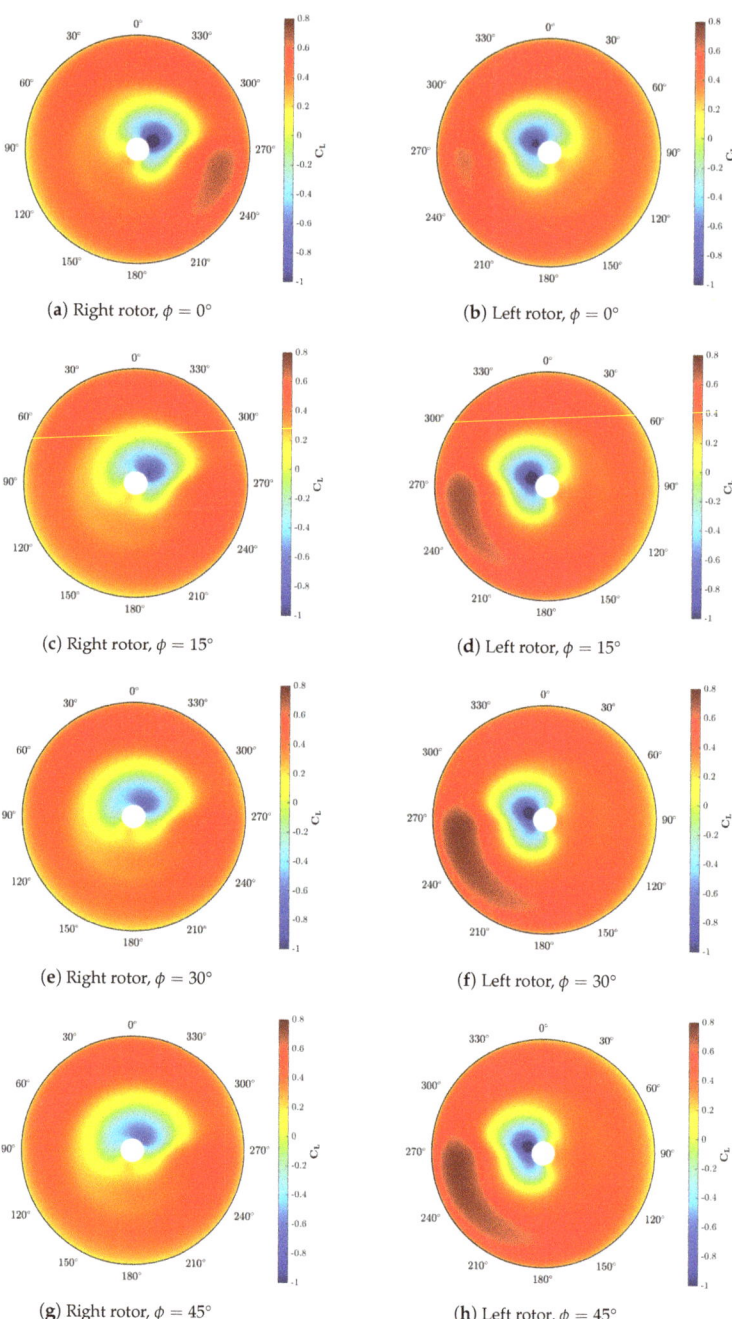

Figure 23. Lift coefficient disks at various bank angles ϕ, front view as shown in Figure 18.

5. Conclusions

A new aeroelastic numerical tool was presented, obtained by coupling a mid-fidelity aerodynamic solver with a multibody dynamics software. It showed the advantages and limits of the proposed approach for the aeroelastic simulation of rotary-wing vehicles. The resulting solver was thoroughly described, highlighting the key-aspects of the mathematical formulation of both solvers, as well as the implementation details of their coupling. Furthermore, the capability to simulate movable surfaces deflection implemented in the aerodynamic solver was highlighted, due to their importance in simulating maneuvers and control studies that represent essential aspects in the design of innovative air vehicles.

The combination of a multibody structural model with a mid-fidelity numerical representation of the aerodynamics enabled to accurately capture aerodynamic interactional effects that characterize rotary-wing aircraft with limited computational cost compared to conventional, higher-fidelity CSD/CFD tools. The coupled code was validated by comparing its results with numerical and experimental data available in the open literature for both fixed- and rotary-wing problems. In detail, validation tests showed the importance of a more accurate description of the aerodynamics to reproduce the dynamic behavior of a wedged wing in flutter condition, and to improve the performance evaluation of a proprotor in hover.

Finally, a complex operating condition was studied by simulating a complete model of the XV-15 tiltrotor performing a roll maneuver. The discussion of the simulation results highlighted the flexibility of the proposed tool in simulating complex rotary-wing aircraft configurations and its capability to capture the fine details of flow physics and the effects on vehicle performance of the aerodynamic interaction between wakes and bodies that are typical of rotary-wing aircraft.

The outcome of this work opens a novel scenario for the scientific and industrial community. Indeed, the significantly lower computational effort required by the coupled simulations with respect to conventional, higher-fidelity CSD tools coupled with URANS solvers, as well as the quite high level of accuracy that could be obtained by these simulations, indicate that the use a mid-fidelity aerodynamic solver for coupled aerodynamics-multibody dynamics simulations represents a valuable instrument for performing the large number of aeroelastic analyses required in the preliminary design phase of innovative rotary-wing vehicles, not only in the rotorcraft field but also in other domains, for example, wind energy and turbomachinery applications.

Author Contributions: Conceptualization, A.S., A.C.; methodology, A.S., A.C., A.Z.; software, A.S., A.C., M.T., P.M.; validation, A.S., A.C.; formal analysis, A.S., A.C.; investigation, A.S., A.C.; resources, A.S., A.C.; data curation, A.S., A.C.; writing—original draft preparation, A.Z., A.S., A.C.; writing—review and editing, A.Z., A.S., A.C., P.M., V.M.; supervision, A.Z., V.M., P.M.; project administration, A.Z., V.M., P.M.; funding acquisition, P.M., V.M. All authors have read and agreed to the published version of the manuscript.

Funding: This research received funding from the Clean Sky 2—H2020 Framework Program, under grant agreement N. 885971 (the FORMOSA project) and under grant agreement N. 863418 (the ATTILA project).

Conflicts of Interest: The authors declare no conflict of interest.

Abbreviations

The following nomenclature and abbreviations are used in this manuscript:

AR	aspect ratio
U_∞	free-stream speed
EA	elastic axis
CG	center of gravity
AoA	angle of attack
f	frequency [Hz]
ζ	damping factor
VL	vortex lattice

SP	surface panel
Ω	rotor speed [RPM]
ρ	air density [kg/m^3]
c	aerodynamic chord [m]
R	rotor radius [m]
N	number of blades
σ	rotor solidity $= (Nc)/(\pi R)$
N/rev	number of rotor revolutions
T	rotor thrust [N]
Q	rotor torque [Nm]
C_T	rotor thrust coefficient $= T/(\rho \pi R^4 \Omega^2)$
C_Q	rotor torque coefficient $= Q/(\rho \pi R^5 \Omega^2)$
FM	rotor figure of merit $= C_T \sqrt{C_T/2}/C_Q$
p	static pressure [Pa]
p_∞	free-stream static pressure [Pa]
C_P	pressure coefficient $= (p - p_\infty)/(0.5 \rho_\infty U_\infty^2)$
C_L	lift coefficient $= L/(0.5 \rho_\infty U_\infty^2 S)$
L	lift [N]
ψ	blade azimuthal angle [deg]
ϕ	bank angle [deg]
$\dot{\phi}$	roll rate [deg/sec]
δ_a	aileron deflection [deg]
b	aircraft span [m]
y	span-wise coordinate along the aircraft span [m]
I_{ij}	Moment of inertia about axis ij [kgm^2]

References

1. Barkai, S.; Rand, O.; Peyran, R.; Carlson, R. Modeling and analysis of tilt-rotor aeromechanical phenomena. *Math. Comput. Model.* **1998**, *27*, 17–43. [CrossRef]
2. Johnson, W. *Camrad II, Comprehensive Analytical Model of Rotorcraft Aerodynamics and Dynamics, Volume I: Theory*; Johnson Aeronautics: Palo Alto, CA, USA, 1992.
3. Bauchau, O.A.; Kang, N.K. A Multibody Formulation for Helicopter Structural Dynamic Analysis. *J. Am. Helicopter Soc.* **1993**, *38*, 3–14. [CrossRef]
4. Yeo, H.; Potsdam, M.; Ortun, B.; Van Truong, K. High-Fidelity Structural Loads Analysis of the ONERA 7A Rotor. *J. Aircr.* **2017**, *54*, 1825–1839. [CrossRef]
5. Ghiringhelli, G.L.; Masarati, P.; Mantegazza, P.; Nixon, M.W. Multi-Body Analysis of a Tiltrotor Configuration. *Nonlinear Dyn.* **1999**, *19*, 333–357. [CrossRef]
6. Quaranta, G.; Masarati, P.; Lanz, M.; Ghiringhelli, G.L.; Mantegazza, P.; Nixon, M.W. Dynamic Stability of Soft-In-Plane Tiltrotors by Parallel Multibody Analysis. In Proceedings of the 26th European Rotorcraft Forum, The Hague, The Netherlands, 26–29 September 2000; pp. 60.1–60.9.
7. Mattaboni, M.; Masarati, P.; Quaranta, G.; Mantegazza, P. Multibody Simulation of Integrated Tiltrotor Flight Mechanics, Aeroelasticity and Control. *J. Guid. Control. Dyn.* **2012**, *35*, 1391–1405. [CrossRef]
8. Masarati, P.; Morandini, M.; Mantegazza, P. An Efficient Formulation for General-Purpose Multibody/Multiphysics Analysis. *J. Comput. Nonlinear Dyn.* **2014**, *9*, 041001. [CrossRef]
9. Cambier, L.; Heib, S.; Plot, S. The Onera elsA CFD software: Input from research and feedback from industry. *Mech. Ind.* **2013**, *14*, 159–174. [CrossRef]
10. Biava, M.; Woodgate, M.; Barakos, G. Fully implicit discrete-adjoint methods for rotorcraft applications. *AIAA J.* **2016**, *54*, 735–749. [CrossRef]
11. Kroll, N.; Eisfeld, B.; Bleecke, H. The Navier-Stokes Code FLOWer. *Notes Numer. Fluid Mech.* **1999**, *71*, 58–71.
12. Biava, M. RANS Computations of Rotor/Fuselage Unsteady Interactional Aerodynamics. Ph.D. Thesis, Politecnico di Milano, Milan, Italy, 2007.
13. Droandi, G.; Zanotti, A.; Gibertini, G. Aerodynamic interaction between rotor and tilting wing in hovering flight condition. *J. Am. Helicopter Soc.* **2015**, *60*, 1–20. [CrossRef]
14. Jimenez Garcia, A.; Barakos, G. CFD Simulations on the ERICA Tiltrotor using HMB2. In Proceedings of the 54th AIAA Aerospace Sciences Meeting, San Diego, CA, USA, 4–8 January 2016.
15. Decours, J.; Beaumier, P.; Khier, W.; Kneisch, T.; Valentini, M.; Vigevano, L. Experimental validation of tilt-rotor aerodynamic predictions. In Proceedings of the 40th European Rotorcraft Forum, Southampton, UK, 2–5 September 2014.
16. Meakin, R. Unsteady Simulation of the Viscous Flow about a V-22 Rotor and Wing in Hover. In Proceedings of the AIAA Atmospheric Flight Mechanics Conference Proceedings, Baltimore, MD, USA, 7–10 August 1995.

17. Potsdam, M.A.; Strawn, R.C. CFD simulations of tiltrotor configurations in hover. *J. Am. Helicopter Soc.* **2005**, *50*, 82–94. [CrossRef]
18. Wissink, A.; Potsdam, M.; Sankaran, V.; Sitaraman, J.; Yang, Z.; Mavriplis, D. Unsteady Simulation of the Viscous Flow about a V-22 Rotor and Wing in Hover. In Proceedings of the American Helicopter Society 66th Annual Forum Proceedings, Phoenix, AZ, USA, 11–13 May 2010.
19. Tran, S.A.; Lim, J.W. Investigation of the Interactional Aerodynamics of the XV-15 Tiltrotor Aircraft. In Proceedings of the 76th Annual Vertical Flight Society Forum and Technology Display, Virtual Event, 5–8 October 2020.
20. Tran, S.; Lim, J.; Nunez, G.; Wissink, A.; Bowen-Davies, G. CFD Calculations of the XV-15 tiltrotor during transition. In Proceedings of the American Helicopter Society 75th Annual Forum, Philadelphia, PA, USA, 13–16 May 2019.
21. Lim, J.W. Fundamental Investigation of Proprotor and Wing Interactions in Tiltrotor Aircraft. In Proceedings of the 75th Annual Vertical Flight Society Forum and Technology Display, Philadelphia, PA, USA, 13–16 May 2019.
22. Tran, S.; Lim, J. Interactional Structural Loads of the XV-15 Rotor in Airplane Mode. In Proceedings of the 45th European Rotorcraft Forum, Warsaw, Poland, 17–20 September 2019.
23. Yeo, H.; Potsdam, M.; Ormiston Robert, A. Rotor Aeroelastic Stability Analysis Using Coupled Computational Fluid Dynamics/Computational Structural Dynamics. *J. Am. Helicopter Soc.* **2011**, *56*, 1–16. [CrossRef]
24. Smith, M.J.; Lim, J.; van der Wall, B.G.; Baeder, J.; Biedron, R.; Boyd, D.; Jayaraman, B.; Jung, S.; Min, B. The HART II international workshop: An assessment of the state of the art in CFD/CSD prediction. *Ceas Aeronaut. J.* **2013**, *4*, 345–372. [CrossRef]
25. Datta, A.; Sitaraman, J.; Chopra, I.; Baeder, J.D. CFD/CSD Prediction of Rotor Vibratory Loads in High-Speed Flight. *J. Aircr.* **2006**, *43*, 1698–1709. [CrossRef]
26. Sitaraman, J.; Roget, B. Prediction of helicopter maneuver loads using a fluid-structure analysis. *J. Aircr.* **2009**, *46*, 1770–1784. [CrossRef]
27. Altmikus, A.; Wagner, S.; Beaumier, P.; Servera, G. A comparison- Weak versus strong modular coupling for trimmed aeroelastic rotor simulations. In Proceedings of the AHS International, 58th Annual Forum Proceedings, Montreal, QC, Canada, 11–13 June 2002; Volume 1, pp. 697–710.
28. Khier, W.; Dietz, M.; Schwarz, T.; Wagner, S. Trimmed CFD Simulation of a Complete Helicopter Configuration. In Proceedings of the 33rd European Rotorcraft Forum, Kazan, Russia, 11–13 September 2007.
29. Guo, T.; Lu, D.; Lu, Z.; Zhou, D.; Lyu, B.; Wu, J. CFD/CSD-based flutter prediction method for experimental models in a transonic wind tunnel with porous wall. *Chin. J. Aeronaut.* **2020**, *33*, 3100–3111. [CrossRef]
30. Massey, S.; Kreshock, R.; Sekula, M. Coupled CFD/CSD Analysis of Rotor Blade Structural Loads with Experimental Validation. In Proceedings of the 31st AIAA Applied Aerodynamics Conference, San Diego, CA, USA, 24–27 June 2013.
31. Winckelmans, G.S. Topics in Vortex Methods for the Computation of Three-and Two-Dimensional Incompressible Unsteady Flows. Ph.D. Thesis, California Institute of Technology, Pasadena, CA, USA, 1989.
32. Cottet, G.H.; Koumoutsakos, P.D. *Vortex Methods: Theory and Practice*; Cambridge University Press: Cambridge, UK, 2000.
33. Lu, Y.; Su, T.; Chen, R.; Li, P.; Wang, Y. A method for optimizing the aerodynamic layout of a helicopter that reduces the effects of aerodynamic interaction. *Aerosp. Sci. Technol.* **2019**, *88*, 73–83. [CrossRef]
34. Alvarez, E.; Ning, A. Development of a Vortex Particle Code for the Modeling of Wake Interaction in Distributed Propulsion. In Proceedings of the AIAA Applied Aerodynamics Conference, Atlanta, GA, USA, 25–29 June 2018.
35. Tan, J.; Zhou, T.; Sun, Y.; Barakos, G. Numerical investigation of the aerodynamic interaction between a tiltrotor and a tandem rotor during shipboard operations. *Aerosp. Sci. Technol.* **2019**, *87*, 62–72. [CrossRef]
36. Tugnoli, M.; Montagnani, D.; Syal, M.; Droandi, G.; Zanotti, A. Mid-fidelity approach to aerodynamic simulations of unconventional VTOL aircraft configurations. *Aerosp. Sci. Technol.* **2021**, *115*, 106804. [CrossRef]
37. Zanotti, A.; Savino, A.; Palazzi, M.; Tugnoli, M.; Muscarello, V. Assessment of a Mid-Fidelity Numerical Approach for the Investigation of Tiltrotor Aerodynamics. *Appl. Sci.* **2021**, *11*, 3385. [CrossRef]
38. Bungartz, H.J.; Lindner, F.; Gatzhammer, B.; Mehl, M.; Scheufele, K.; Shukaev, A.; Uekermann, B. preCICE—A fully parallel library for multi-physics surface coupling. *Comput. Fluids* **2016**, *141*, 250–258. [CrossRef]
39. Cocco, A.; Savino, A.; Zanotti, A.; Zanoni, A.; Masarati, P.; Muscarello, V. Coupled Multibody-Mid Fidelity Aerodynamic Solver for Tiltrotor Aeroelastic Simulation. In Proceedings of the 9th International Conference on Computational Methods for Coupled Problems in Science and Engineering, Virtual Event, 13–16 June 2021.
40. Savino, A.; Cocco, A.; Zanoni, A.; Zanotti, A.; Muscarello, V. A Coupled Multibody–Mid Fidelity Aerodynamic Tool for the Simulation of Tiltrotor Manoeuvres. In Proceedings of the 47th European Rotorcraft Forum, Virtual Event, 7–10 September 2021.
41. Maisel, M. *NASA/Army XV-15 Tilt-Rotor Research Aircraft Familiarization Document*; TM X-62,407; NASA: Washington, DC, USA, 1975.
42. Churchill, G.; Gerdes, R. *Advanced AFCS Developments on the XV-15 Tilt Rotor Research Aircraft. [Automatic Flight Control System]*; NASA Technical Report; NASA: Washington, DC, USA, 1984.
43. Masarati, P.; Morandini, M. Intrinsic Deformable Joints. *Multibody Syst. Dyn.* **2010**, *23*, 361–386. [CrossRef]
44. Ghiringhelli, G.L.; Masarati, P.; Mantegazza, P. A Multi-Body Implementation of Finite Volume C^0 Beams. *AIAA J.* **2000**, *38*, 131–138. [CrossRef]
45. Bauchau, O.A.; Betsch, P.; Cardona, A.; Gerstmayr, J.; Jonker, B.; Masarati, P.; Sonneville, V. Validation of flexible multibody dynamics beam formulations using benchmark problems. *Multibody Syst. Dyn.* **2016**, *37*, 29–48. [CrossRef]

46. Morandini, M.; Masarati, P. Implementation and Validation of a 4-Node Shell Finite Element. In Proceedings of the International Design Engineering Technical Conferences and Computers and Information in Engineering Conference, Buffalo, NY, USA, 17–20 August 2014.
47. Craig, R.R., Jr.; Bampton, M.C.C. Coupling of Substructures for Dynamic Analysis. *AIAA J.* **1968**, *6*, 1313–1319. [CrossRef]
48. Masarati, P.; Sitaraman, J. Tightly Coupled CFD/Multibody Analysis of NREL Unsteady Aerodynamic Experiment Phase VI Rotor. In Proceedings of the 49th AIAA Aerospace Sciences Meeting, Orlando, FL, USA, 4–7 January 2011.
49. Piszkin, S.T.; Levinsky, E. *Nonlinear Lifting Line Theory for Predicting Stalling Instabilities on Wings of Moderate Aspect Ratio*; Technical Report; General Dynamics San Diego CA Convair Div: San Diego, CA, USA, 1976.
50. Gallay, S.; Laurendeau, E. Nonlinear generalized lifting-line coupling algorithms for pre/poststall flows. *AIAA J.* **2015**, *53*, 1784–1792. [CrossRef]
51. Morino, L.; Kuot, C.C. Subsonic potential aerodynamics for complex configurations: A general theory. *AIAA J.* **1974**, *12*, 191–197. [CrossRef]
52. Katz, J.; Plotkin, A. *Low-Speed Aerodynamics*; Cambridge University Press: Cambridge, UK, 2001; Volume 13.
53. Leishman, J.G. *Principles of Helicopter Aerodynamics*; Cambridge University Press: Cambridge, UK, 2000.
54. Lindsay, K.; Krasny, R. A particle method and adaptive treecode for vortex sheet motion in three-dimensional flow. *J. Comput. Phys.* **2001**, *172*, 879–907. [CrossRef]
55. Rendall, T.C.; Allen, C.B. Unified fluid–structure interpolation and mesh motion using radial basis functions. *Int. J. Numer. Methods Eng.* **2008**, *74*, 1519–1559. [CrossRef]
56. Goland, M. The flutter of a uniform cantilever wing. *J. Appl. Mech.-Trans. ASME* **1945**, *12*, A197–A208. [CrossRef]
57. Patil, M.J.; Hodges, D.H.; Cesnik, C.E. Nonlinear aeroelastic analysis of complete aircraft in subsonic flow. *J. Aircr.* **2000**, *37*, 753–760. [CrossRef]
58. Murua, J.; Palacios, R.; Graham, J.M.R. Assessment of wake-tail interference effects on the dynamics of flexible aircraft. *AIAA J.* **2012**, *50*, 1575–1585. [CrossRef]
59. Hua, Y.; Sarkar, T.K. Matrix pencil method for estimating parameters of exponentially damped/undamped sinusoids in noise. *IEEE Trans. Acoust. Speech Signal Process.* **1990**, *38*, 814–824. [CrossRef]
60. Wang, Z.; Chen, P.; Liu, D.; Mook, D.; Patil, M. Time domain nonlinear aeroelastic analysis for HALE wings. In Proceedings of the 47th AIAA/ASME/ASCE/AHS/ASC Structures, Structural Dynamics, and Materials Conference, 14th AIAA/ASME/AHS Adaptive Structures Conference, Newport, RI, USA, 1–4 May 2006.
61. Bilger, J.; Marr, R.; Zahedi, A. Results of Structural Dynamic Testing of the XV-15 Tilt Rotor Research Aircraft. *J. Am. Helicopter Soc.* **1982**, *27*, 58–65. [CrossRef]
62. Acree, C.W.; Peyran, R.J.; Johnson, W. Rotor Design Options for Improving Tiltrotor Whirl-Flutter Stability Margins. *J. Am. Helicopter Soc.* **2001**, *46*, 87–95. [CrossRef]
63. Acree, C., Jr. *An Improved CAMRAD Model for Aeroelastic Stability Analysis of the XV-15 with Advanced Technology Blades*; NASA Technical Memorandum 4448; NASA: Washington, DC, USA, 1993.
64. Felker, F.F.; Betzina, M.D. *Performance and Loads Data from a Hover Test of a Full-Scale XV-15 Rotor*; Technical Report; NASA Ames Research Center: Moffett Field, CA, USA, 1985.
65. Masarati, P.; Morandini, M. An Ideal Homokinetic Joint Formulation for General-Purpose Multibody Real-Time Simulation. *Multibody Syst. Dyn.* **2008**, *20*, 251–270. [CrossRef]
66. Leishman, J.G. *Principles of Helicopter Aerodynamics*, 2nd ed.; Cambridge University Press: Cambridge, UK, 2006.
67. Montagnani, D.; Tugnoli, M.; Fonte, F.; Zanotti, A.; Syal, M.; Droandi, G. Mid–fidelity analysis of unsteady interactional aerodynamics of complex vtol configurations. In Proceedings of the 45th European Rotorcraft Forum (ERF 2019), Warsaw, Poland, 17–20 September 2019; pp. 1–11.
68. Felker, F.F.; Young, L.A.; Signor, D.B. *Performance and Loads Data from a Hover Test of a Full-Scale Advanced Technology XV-15 Rotor*; Technical Memorandum 86854; NASA Ames Research Center: Mountain View, CA, USA, 1986.
69. Drela, M. XFOIL: An Analysis and Design System for Low Reynolds Number Airfoils. In *Low Reynolds Number Aerodynamics*; Mueller, T.J., Ed.; Springer: Berlin/Heidelberg, Germany, 1989; pp. 1–12.
70. Viterna, L.A.; Janetzke, D.C. *Theoretical and Experimental Power from Large Horizontal-Axis Wind Turbines*; Technical Report; Washington Procurement Operations Office: Washington, DC, USA, 1982.
71. Ferguson, S.W. *Development and Validation of a Simulation for a Generic Tilt-Rotor Aircraft*; CR 166537; NASA: Washington, DC, USA, 1989.

Article

An Experimental-Numerical Investigation of the Wake Structure of a Hovering Rotor by PIV Combined with a Γ_2 Vortex Detection Criterion

Fabrizio De Gregorio, Antonio Visingardi and Gaetano Iuso

1. Italian Aerospace Research Centre—CIRA, Aerodynamic Measurement Methodologies Laboratory, 81043 Capua, Italy
2. Italian Aerospace Research Centre—CIRA, Fluid Mechanics Laboratory, 81043 Capua, Italy; a.visingardi@cira.it
3. Politecnico di Torino, Mechanical and Aerospace Engineering Department, 10129 Turin, Italy; gaetano.iuso@polito.it
* Correspondence: f.degregorio@cira.it

Abstract: The rotor wake aerodynamic characterization is a fundamental aspect for the development and optimization of future rotary-wing aircraft. The paper is aimed at experimentally and numerically characterizing the blade tip vortices of a small-scale four-bladed isolated rotor in hover conditions. The investigation of the vortex decay process during the downstream convection of the wake is addressed. Two-component PIV measurements were carried out below the rotor disk down to a distance of one rotor radius. The numerical simulations were aimed at assessing the modelling capabilities and the accuracy of a free-wake Boundary Element Methodology (BEM). The experimental and numerical results were investigated by the Γ_2 criterion to detect the vortex location. The rotor wake mean velocity field and the instantaneous vortex characteristics were investigated. The experimental/numerical comparisons show a reasonable agreement in the estimation of the mean velocity inside the rotor wake, whereas the BEM predictions underestimate the diffusion effects. The numerical simulations provide a clear picture of the filament vortex trajectory interested in complex interactions starting at about a distance of $z/R = -0.5$. The time evolution of the tip vortices was investigated in terms of net circulation and swirl velocity. The PIV tip vortex characteristics show a linear mild decay up to the region interested by vortex pairing and coalescence, where a sudden decrease, characterised by a large data scattering, occurs. The numerical modelling predicts a hyperbolic decay of the swirl velocity down to $z/R = -0.4$ followed by an almost constant decay. Instead, the calculated net circulation shows a gradual decrease throughout the whole wake development. The comparisons show discrepancies in the region immediately downstream the rotor disk but significant similarities beyond $z/R = -0.5$.

Keywords: rotary-wing aerodynamics; vortex detection criterion; BEM method; tip vortex interactions

Citation: De Gregorio, F.; Visingardi, A.; Iuso, G. An Experimental-Numerical Investigation of the Wake Structure of a Hovering Rotor by PIV Combined with a Γ_2 Vortex Detection Criterion. *Energies* **2021**, *14*, 2613. https://doi.org/10.3390/en14092613

Academic Editors: Oriol Lehmkuhl and Adrian Ilinca

Received: 2 March 2021
Accepted: 28 April 2021
Published: 2 May 2021

Publisher's Note: MDPI stays neutral with regard to jurisdictional claims in published maps and institutional affiliations.

Copyright: © 2021 by the authors. Licensee MDPI, Basel, Switzerland. This article is an open access article distributed under the terms and conditions of the Creative Commons Attribution (CC BY) license (https://creativecommons.org/licenses/by/4.0/).

1. Introduction

During the generation of the required thrust, the helicopter rotor blades produce a complex wake system which is characterized by spanwise shed vortices and trailing vortices having a strength varying along the blade span. In particular, a helical system of strong blade tip vortices develops because of the rotor revolution This vortex system can interact with the main rotor, the tail rotor, and the airframe. Hence, vortex formation and development are important factors influencing the aerodynamics of the entire helicopter. In undisturbed hover, the blade pitch setting does not vary with time and constant lift and induced velocity distributions are therefore generated during the rotor revolution. This leads to a relatively simple vortex system characterized by vortices keeping a constant strength over the azimuth. The characteristics and development of the tip vortices are aspects of fundamental importance for the understanding of the rotor wake evolution.

Over the past twenty years, Particle Image Velocimetry (PIV) has made the detailed investigation of the properties of the blade tip vortices possible by providing planar velocity data. Many PIV measurements on hovering helicopter rotors have been performed. Martin et al. [1] applied phase-resolved stereoscopic PIV measuring on a sub-scale two-bladed helicopter rotor in hovering up to one revolution age. The results were compared with three-component Laser Doppler Velocimetry (LDV) measurements using the same seeding medium. Information on the required spatial resolution to resolve the tip vortices was provided. In the same year, Heineck et al. [2] used a stereo PIV set-up on a two-bladed rotor test stand to capture blade tip vortices up to an age of 270 deg, focusing on the data processing and the effect of the vortex wandering on the vortex core size. McAlister [3] investigated the rotor wake of a two-bladed model helicopter up to a wake age of 390 deg. Conclusions on the temporal development of the maximum swirl velocity as well as the core growth were drawn using a stereo PIV setup. Richard and van der Wall [4] published an analysis of the two-component and stereo PIV on a four-bladed rotor in hover condition up to the half revolution, which encompassed a three-dimensional reconstruction of a vortex filament based on the PIV data and indication about the necessary spatial resolution to resolve properly the vortex characteristics. Many studies have seen the use of PIV for the investigation of the rotor aerodynamics or the full helicopter configuration [5] and a complete overview is given by the paper of Raffel et al. [6].

Several methodologies have been developed and applied during the years to the numerical investigation of the helicopter aerodynamics and the related wake structure, each with their level of sophistication and limitations. A historical review of these methodologies can be found in Leishman [7] and Johnson [8]. Two examples can be mentioned to highlight the pros and cons in the application of these methodologies: considering the sophisticated CFD-based tools, they solve the governing fluid dynamic equations in a region of space surrounding the configuration to be analysed with a level of accuracy that is dependent on the resolution of the volume grid applied for the numerical discretization, and this requires to be extremely fine around the blades and/or the airframe surfaces and in the wake downstream, where the viscous stresses are the highest, to avoid an excessive non-physical numerical dissipation. For this reason, these methods are usually computationally onerous. In addition, the accuracy of the solution also depends on the turbulence model applied. Conversely, considering the lower-fidelity free-wake panel methods, they require limited computational resources because are based on the simplifying assumptions of inviscid and incompressible flow and need just a surface discretization of the wake and body geometries. They are free from numerical dissipation but need the use of suitable regularization models to avoid unphysical singularities produced by application of the Biot–Savart law for the evaluation of the velocities induced by the wake vortex filaments and appropriate models to model the wake dissipation process caused by the natural flow viscosity.

Regardless of the applied methodology, it appears that the accurate numerical reproduction of blade tip vortices is challenging and requires validation [9]. Duraisamy et al. [10] published a direct comparison of a computational fluid dynamics (CFD) simulation and respective experiments focusing on the physics of vortex formation from a single-bladed hovering rotor. They found PIV to be a valid method for a qualitative and quantitative comparison. Unfortunately, the amount of experimental data concerning the vortex formation, development, and decay for the validation of the numerical approaches in rotating systems mostly do not investigate beyond one rotor revolution. In the past, Caradonna [11] investigated the tip trajectory up to a wake age of 1080 deg. by flow visualization and these data were used for validation [12,13]. More recently, the large interest in the brownout phenomenon drove many investigations to characterise the rotor wake. Lee [14] presented remarkable flow visualization results of a hovering rotor out of ground effect but without providing quantitative data of the blade tip vortices evolution.

The current work stems from research activity in the framework of the GARTEUR Action Group 22 Forces on Obstacles in Rotor Wake (Visingardi et al. [15]) to evaluate the mutual effects numerically and experimentally between a small-scale helicopter rotor in hover flight and a cylindrical sling load located at different positions below the rotor disk. The experimental investigations focused on the development of the rotor wake far from the rotor disk. The numerical simulations were performed with the main purpose to assess the modelling capabilities and the accuracy of a free-wake Boundary Element Methodology. The results were presented in a paper of the 44th European Rotorcraft Forum discussing the effect of the wake on the sling load [16].

Following this activity, new investigations focused on the blade tip vortex characteristics and the dissipation phenomena when moving away from the disk of the isolated rotor (i.e., without cylinder) up to a wake age of three rotor revolutions. Particular attention was dedicated to the vortex detection criterion adopted on the PIV data. The most widely used local methods for vortex detection are founded on the velocity gradient tensor ∇u and its three invariants. Examples are the Δ criterion introduced by Dallmann [17], Vollmers et al. [18], and Chong et al. [19]; the Q criterion by Hunt et al. [20]; λ_2 criterion by Jeong and Hussain [21]. These local vortex-detection criteria are not always suitable for noisy PIV data affected by spurious vectors, thus resulting in high-velocity gradients. The Γ_2 criterion proposed by Graftieaux et al. [22] offered a possible solution and later it was successfully applied to complex wind tunnel measurements by Mulleners and Raffel [23].

The current work illustrates the results of these new numerical and experimental investigations. The paper is organized into sections. Section 2 describes the main characteristics of the four-bladed rotor rig, the PIV system including the data evaluation procedure and a description of the adopted vortex identification criterion. The numerical methodology is illustrated in Section 3, whereas the numerical/experimental comparison of the results obtained is fully documented in Section 4. The conclusions are finally reported in Section 5.

2. Experimental Setup and Test Conditions

A dedicated rotor test rig was built based on an existing commercial radio-controlled helicopter model (Blade 450 3D RTF), (Figure 1a). The original two-bladed rotor was replaced by a four-bladed one with collective and cyclic control. The rotor blades were untwisted, with rectangular planform and parabolic tip. The blades presented a radius of $R = 0.36$ m and a chord length of $c = 0.0327$ m. The root cut-out was located at 16% of the radius. A NACA0013 airfoil was used throughout the blade span. The blade planform and the geometry of the airfoil are shown in Figure 1b. The rotor solidity was equal to $\sigma = (N_b c)/\pi R = 0.116$ and the aspect ratio of the blades was $AR = R/c \approx 11$. The rotor spun clockwise when seen from above, the collective pitch angle θ_0 varied from $1°$ to $12.2°$ and the maximum speed was $\Omega = 1780$ RPM.

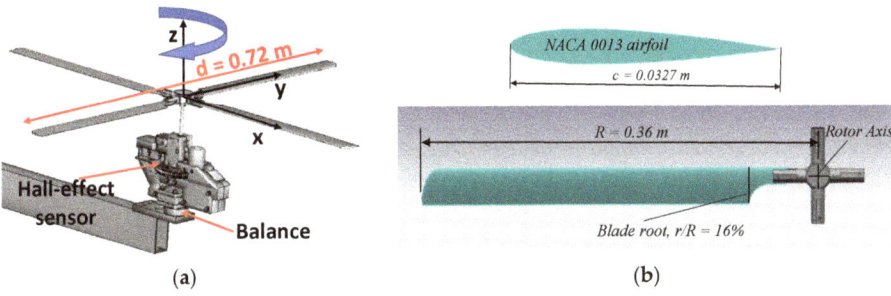

Figure 1. (a) Rotor test rig; (b) airfoil and planform blade drawings, (reprinted from ref. [24]).

The force and moments produced by the rotor rig were measured by a six components balance (ATI MINI40). The detailed characteristics in terms of full scale and accuracy are summarized in Table 1.

Table 1. Balance characteristics.

	Fx (N)	Fy (N)	Fz (N)	Mx (Nm)	My (Nm)	Mz (Nm)
Full scale	±20	±20	±60	±1	±1	±1
Accuracy (%FS)	0.25	0.25	0.60	0.0125	0.0125	0.0125

The rotor test rig was located at a distance of 5 R from the floor and 3 R from the ceiling to avoid any influence of the surrounding walls. One hall-effect sensor was located on the shaft gear for measuring the rotating speed and for providing a trigger TTL signal at a prefixed azimuth angle to allow phase-locked measurements.

The current investigation addressed the wake downwash generated by the four-bladed rotor in hover conditions at a constant collective angle of $\theta_0 = 11.8°$. The angular velocity was set to Ω = 1740 RPM, which leads to a blade tip velocity of V_{tip} = 66 m/s and a thrust value of T = 12 N. The resulting blade loading was $C_T/\sigma = T/\left(\rho A \Omega^2 R^2 \sigma\right) = 0.052$ with the density ρ = 1.114 kg/m^3, and the rotor area A = 0.41 m^2. The Mach and Reynolds numbers are given at the radius tip (Table 2).

Table 2. Parameter of the measured test case.

$\Omega/2\pi$ (Hz)	V_{tip} (m/s)	M_{tip}	Re_{tip}
29	65.6	0.19	1.47*10^5

2.1. PIV System and Evaluation

A fixed frame of reference was defined and having the origin located in the rotor centre with the *x*-axis horizontally oriented along the rotor blade, the *y*-axis orthogonal to the *x*-axis and lying in the rotor disk plane, and the *z*-axis vertically and upward directed. The rotor wake characteristics were investigated by a standard two-component PIV measurement system composed of a double head Nd-Yag laser with a maximum energy of 300 mJ per pulse at 532 nm and a single double frame CCD camera (2048 by 2048 px) with a dynamic range of 14 bits. The camera was mounted on two components translating system to cover the full region of interest in the xz-plane. The light sheet was vertically oriented and aligned with the rotor blade at the azimuth angle of Ψ = 180°. A standard PIV layout was adopted with the camera line of sight orthogonal to the laser sheet (Figure 2).

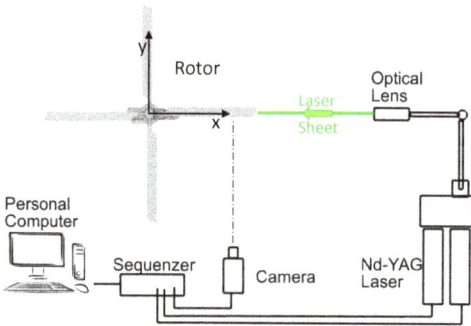

Figure 2. PIV layout: plan-view.

The wake downwash was investigated up to 2.25 rotor radii from the rotor disk. The camera, equipped with a 50 mm lens, recorded a measurement region of 320 mm by 320 mm, with a spatial resolution of 6 pixel/mm in the image plane. Separate measurement regions needed to cover the full wake (Figure 3a). The data post processing discussed in [16] provided a spatial resolution of $\Delta x = 2.5$ mm not sufficient for the characterization of the tip vortices.

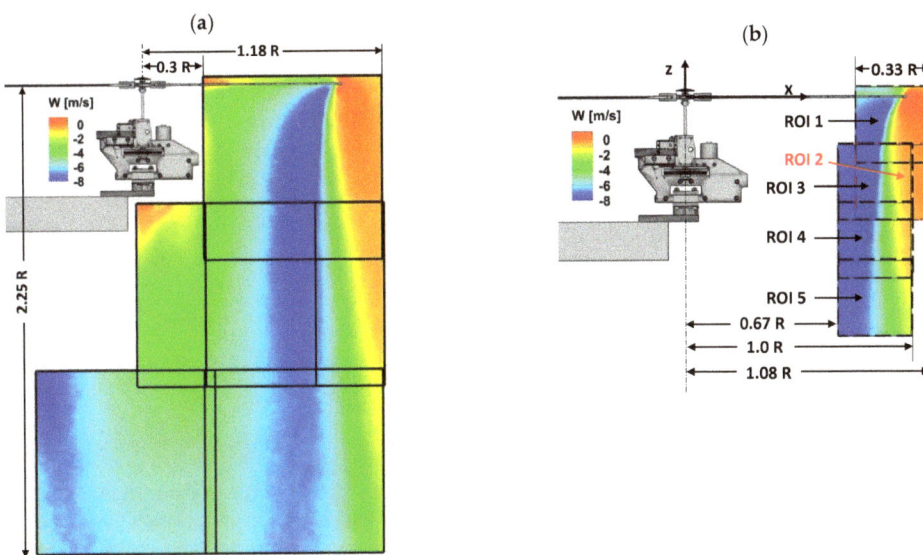

Figure 3. PIV measurement regions: (**a**) Rotor wake measurements, (**b**) Tip vortex characterization.

To track the blade tip vortices in proximity to the rotor disk, the camera was equipped with a lens featuring a fixed focal length of f = 200 mm, and the f-number was set to $f_\# = 2.8$. The measurement region was located immediately below the rotor disk plane on a vertical plane radially ranging between x/R = 0.68 and x/R = 1.08 to identify the trajectories of the trailing tip vortices. Five PIV measurement regions with the size of about 120 mm by 120 mm, partially overlapped, were used to measure the rotor wake, and in particular the blade tip vortices characteristics up to one radius downstream the rotor disc, Figure 3b. This yielded a spatial resolution of about 17 px/mm in the image plane. The delay time between the two laser pulses was set to 25 µs according to the highest expected velocities in the flow. As tracer particles, sprayed diethylhexylsebacate (DEHS) oil was used. A seeding generator with 20 Laskin nozzles provided oil droplets with an average size less than 1 µm. The full test room was seeded to have a homogenous concentration of particles. More than 150 image pairs were recorded at a frame rate of 3 Hz for each region of interest (ROI) over about 1450 rotor revolutions. The PIV images were pre-processed by applying a background grey-level subtraction. PIV-View 3.60 was used to process the images. The analysis consisted in a Multi-grid scheme with a B-Spline of 3rd order image deformation ending at 32×32 px^2 and 75% overlap. Correlation maps were calculated by FFT multiple correlations (Hart [25]) of 2 windows and a 3-point Gaussian peak fit was used to obtain the displacement.

The results presented a velocity spatial resolution of $\Delta x = 0.47$ mm. The random noise of the PIV cross-correlation procedure can be estimated as 0.1 px as a rule–of–thumb (Raffel et al. [26]). Using the current values for the optical resolution (17 px/mm) and the laser double–pulse delay (25 µs), this leads to a velocity error of ΔV of ≈ 0.23 m/s for the PIV measurements. In the proximity to the rotor disk, the core radius r_c of the tip vortices was measured. The core radius is defined as the distance from the vortex centre to the radial position where the maximum swirl velocity is reached (Figure 4). Values of r_c between 3 and 3.3 mm were measured giving a ratio $\Delta x/r_c$ of about 0.15–0.14 according to the value of $\Delta x/r_c \leq 0.2$ indicated by Martin et al. [1] to guarantee a correct vortex characterization.

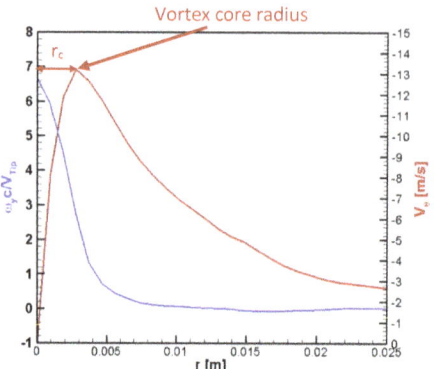

Figure 4. Tip vortex main characteristics. Tangential velocity (red curve) and normalised out-of-plane vorticity (blue curve) vs. vortex radius. The vortex core radius is detected by the maximum tangential velocity (V_θ).

2.2. Vortex-Identification Criterion

To overcome the difficulty in identifying the centre of the vortices in the presence of spurious vectors, a method based on the velocity field topology, without using velocity derivatives was chosen. This was the Γ_2-criterion proposed by Graftieaux et al. [22]. The function Γ_2 is defined in discrete form as:

$$\Gamma_2\left(\vec{x_i}\right) = \frac{1}{M} \sum_{x_j \in S_i} \frac{\left\{\left(\vec{x_j} - \vec{x_i}\right) \times \left(\vec{u_j} - \vec{u}_{mean}\right)\right\} \cdot \vec{n}}{\left|\vec{x_j} - \vec{x_i}\right| \left|\left(\vec{u_j} - \vec{u}_{mean}\right)\right|} \quad (1)$$

with S_i representing a two-dimensional circle around x_i with radius D, M the number of grid points x_j inside S_i with $j \neq i$, u_{mean} the average velocity vector within the region S, $\vec{n}$ the unit vector normal to the PIV plane and u_j the velocity at x_j. The l radius D of the domain S_i is expressed in terms of grid spacing. Γ_2 is a 3D non-dimensional scalar function, with $-1 \leq \Gamma_2 \leq 1$. The zones delimited by $|\Gamma_2| > \frac{2}{\pi}$ identify the vortices present in the measurement region. The vortex centre is identified as the maximum value of the absolute of Γ_2 in the delimited zone. In cases of PIV data affected by spurious vectors or data void, it is suggested to use the weighted centroid of the Γ_2 for the identification of the centre of the vortex. In the current work, the weighted centroid is chosen to identify the centres. The choice of the domain radius D influences the accuracy of the centre detection and the dimension of the identified vortices. De Gregorio and Visingardi [27] indicated that the selection of a domain radius D, equal to the larger core radius existing within the investigated flow field, provides the best measurement accuracy for single and multiple vortices affected by spurious vectors. In the case of elliptical vortices, a domain radius D equal to the semi-major axis is recommended for reducing centre detection errors, whereas the magnitude of the domain radius is recommended to be set between two to six times the core radius size in the case of significant data void in the vortex core.

In the processing of the PIV images, the lack of particles caused a number of spurious velocity vectors. A normalised median test, presented by Westerweel and Scarano [28], coupled to a Dynamic Mean Test and a Global Histogram Filter (described by Raffel et al. [26]) were used to detect and then remove the spurious vectors. After deleting the outliers, the missing data were replaced using bi-linear interpolation. Taking into account the particle void, the domain radius D was fixed equal to 10 grid spacing, which was almost twice the value of the vortex core radius ($r_c \approx 6$ grid point as shown in Figure 5). Once the centre was detected, the vortex characteristics were calculated on concentric circles moving away from the centre. The raw or filtered data have shown scattering data in the core, while the interpolated data provided a satisfactory agreement with the Vatistas [29] theoretical curve (Figure 5). The agreement between the linear interpolated data and the Vatistas vortex is explained by the nature of a real vortex core where the tangential velocity behaviour is linear and can be expressed as $V_\theta = \Omega \cdot r$, thus justifying the application of a bi-linear interpolation to account for the missing vectors.

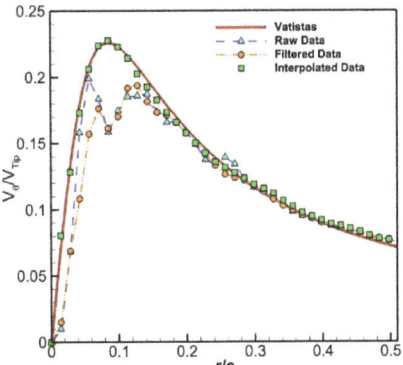

Figure 5. Vortex characteristics: swirl velocity for raw, filtered and interpolated vs. Vatistas.

3. Numerical Methodology

The numerical simulations were carried out by using the code RAMSYS [30], which is an unsteady, inviscid and incompressible free-wake vortex lattice Boundary Element Method (BEM) solver for multi-rotor, multi-body configurations developed at CIRA. It is based on Morino's boundary integral formulation for the solution of the Laplace equation for the velocity potential φ [31].

3.1. Governing Equation

The fluid is assumed to be inviscid, incompressible and irrotational. Under these hypotheses there exists a function $\varphi(\mathbf{x}, t)$, called velocity potential, such that the velocity vector $\mathbf{v} = \nabla \varphi$ and the continuity equation reduces to Laplace equation:

$$\nabla^2 \varphi(\mathbf{x}, t) = 0 \qquad (2)$$

The application of the Green's function to Equation (2) yields a boundary integral representation of the velocity potential:

$$\varphi(\mathbf{x}, t) = \int_{S_B} \left(G \frac{\partial \varphi}{\partial n} - \varphi \frac{\partial G}{\partial n} \right) dS(\mathbf{y}) - \int_{S_W} \Delta \varphi \frac{\partial G}{\partial n} dS(\mathbf{y}) \qquad (3)$$

where S_B and S_W are body and wake surfaces, respectively, and $G = -1/4\pi ||\mathbf{y} - \mathbf{x}||$ is the free-space fundamental solution of the 3D Laplace equation.

3.2. Boundary Conditions

Far from the body, the velocity potential is null at infinity:

$$\lim_{\mathbf{x} \to \infty} \varphi(\mathbf{x}, t) = 0 \qquad (4)$$

The impermeability boundary condition on S_B yields:

$$\frac{\partial \varphi}{\partial n} = \mathbf{V}_B \cdot \mathbf{n} \tag{5}$$

where $\mathbf{V}_B$ denotes the velocity of body points and $\mathbf{n}$ denotes the outward unit normal vector on S_B.

The boundary condition on the wake is expressed as:

$$\frac{D_W \Delta \varphi}{Dt} = 0 \tag{6}$$

according to which the potential jump $\Delta \varphi$ across the wake surface is constant:

$$\Delta \varphi(\mathbf{x_W}, t) = \Delta \varphi^{TE}(t - \tau) \tag{7}$$

with τ denoting the time taken by the wake material point $\mathbf{x_W}$ to move from the trailing edge to its current position.

3.3. Novel Boundary Integral Formulation

The novel formulation proposed by Gennaretti et al. [32] is applied in RAMSYS to avoid the instabilities arising in the numerical formulation when wake panels are too close to or impinge the body. In this formulation, the velocity potential φ is split into a scattered potential φ_s, generated by sources and doublets over S_B and doublets over the part of the wake surface in contact with the blade trailing edge (Near wake, S_W^N), and an incident potential φ_I, generated by doublets over the part of the wake not in contact with the trailing edge (Far wake, S_W^F), such that $S_W = S_W^N \cup S_W^F$ and $\varphi = \varphi_s + \varphi_I$.

The application of this novel formulation provides a new expression of Equation (3), which is then replaced by:

$$\varphi_s(\mathbf{x}, t) = \int_{S_B} \left[G(\chi - \chi_I) - \varphi_s \frac{\partial G}{\partial n} \right] dS(\mathbf{y}) - \int_{S_W^N} \Delta \varphi_s^{TE}(t - \tau) \frac{\partial G}{\partial n} dS(\mathbf{y}) \tag{8}$$

and by:

$$\varphi_I(\mathbf{x}, t) = -\int_{S_W^F} \Delta \varphi_I \frac{\partial G}{\partial n} dS(\mathbf{y}) \tag{9}$$

where the boundary conditions are given by:

$$\frac{\partial \varphi_s}{\partial n} = \mathbf{V}_B \cdot \mathbf{n} - \mathbf{v}_I \cdot \mathbf{n} = \chi - \chi_I \tag{10}$$

and:

$$\Delta \varphi_I(\mathbf{x_W}, t) = \Delta \varphi_s^{TE}(t - \tau) \tag{11}$$

with the velocity induced by the far wake obtained from the gradient of Equation (9) combined with Equation (11),

$$\mathbf{v}_I(\mathbf{x}, t) = -\nabla_\mathbf{x} \int_{S_W^F} \Delta \varphi_s^{TE}(t - \tau) \frac{\partial G}{\partial n} dS(\mathbf{y}) \tag{12}$$

3.4. Numerical Solution

The numerical solution of Equation (8) is obtained by defining boundary elements, i.e., by discretizing S_B and S_W into quadrilateral panels, assuming φ_s, $\partial \varphi_s / \partial n$, $\Delta \varphi_s$ to be piecewise constant, and imposing that the equation is satisfied at the centre of each body element (collocation method). Specifically, dividing the blade surface into M panels, S_{B_i}, and the wake surface into N panels, S_{W_j}, the discretized version of Equation (8) gives the linear algebraic system:

$$\varphi_{sk} = \sum_{i=1}^{M} B_{ki}[\chi(t) - \chi_I(t)] + \sum_{i=1}^{M} C_{ki}\varphi_{si}(t) + \sum_{j=1}^{N} F_{kj}\Delta\varphi_{sj}(t) \quad (13)$$

where $\varphi_{si}(t) = \varphi_s(\mathbf{x}_i, t)$, $\chi_i(t) = \chi(\mathbf{x}_i, t)$, $\chi_{Ii}(t) = \chi_I(\mathbf{x}_i, t)$, $\Delta\varphi_{sj}(t) = \Delta\varphi_{sj}^{TE}(t)$ whereas source/sink and doublet coefficients are given by:

$$B_{ki} = -\frac{1}{4\pi}\int_{S_{B_i}}\left(\frac{1}{|\mathbf{y}-\mathbf{x}_k|}\right)dS \quad C_{ki} = -\frac{1}{4\pi}\int_{S_{B_i}}\frac{\partial}{\partial n}\left(\frac{1}{|\mathbf{y}-\mathbf{x}_k|}\right)dS \quad F_{kj} = -\frac{1}{4\pi}\int_{S_{W_j}}\frac{\partial}{\partial n}\left(\frac{1}{|\mathbf{y}-\mathbf{x}_k|}\right)dS \quad (14)$$

The solution of the algebraic system is obtained by the application of the GMRES iterative method.

Once the potential field is known, the surface pressure distributions are evaluated by applying the unsteady version of the Bernoulli equation:

$$\frac{\partial\varphi_s}{\partial t} + \frac{\partial\varphi_I}{\partial t} - \mathbf{V}_B\cdot(\nabla\varphi_s + \mathbf{v}_I) + \frac{||\nabla\varphi_s + \mathbf{v}_I||^2}{2} + \frac{P}{\rho} = \frac{P_\infty}{\rho} \quad (15)$$

where the incident potential φ_I is obtained from integration of the incident velocity field by using Equation (12) with the inclusion of the vortex-core model. Finally, the evaluation of the forces and moments is obtained by the integration of Equation (15).

3.5. Vortex Core Model

To account for the viscous diffusion of the wake vortex elements, the Vatistas vortex core model was used, according to which the swirl velocity is expressed as:

$$V_\theta = \frac{r\Gamma_v}{2\pi(r^{2n} + r_c^{2n})^{\frac{1}{n}}} \quad (16)$$

where the coefficient n has been set equal to "1", as suggested by Scully [33].

The applied diffusion model is the one described by Squire [34]. In this model, the growth with the time of the core radius r_c is given by:

$$r_c = \sqrt{r_{c0}^2 + 4\alpha\delta\nu t} \quad (17)$$

where the term r_{c0} is the initial core radius that removes the singularity at t_0, and was set equal to 5% of the blade average chord length c in the calculations, the term α is the Oseen coefficient and is equal to 1.25643. The product $\delta\nu$ represents the "eddy viscosity" where ν is the kinematic viscosity and:

$$\delta = 1 + a_1\frac{\Gamma_v}{\nu} \quad (18)$$

represents an average effective (turbulent) viscosity coefficient in which Γ_v is the circulation strength of the vortex element, while the Squire's coefficient a_1 is an empirical parameter specified to vary between 0.2 and 0.0002, as indicated in Bhagwat [35]. For a small-scale rotor, like the one used for these investigations, a value of $O(10^{-4})$ can be used. The model suggested by Donaldson & Bilanin [36] was used to take into account the decay of the circulation Γ_v with time. According to this model, the circulation of the tip vortex $\Gamma_v(t)$ is expressed as:

$$\Gamma_v(t) = \Gamma_0\exp\left(-\frac{bq}{s}t\right) \quad (19)$$

being b a decay coefficient; q the ambient turbulence level and s the aircraft semispan. In the present calculations, the coefficient bq/s was replaced with a single coefficient set equal to 2.5. This value was tuned, together with the empirical parameter a_1, finally set to 0.0003, to match several experimental observations [35,37,38] according to which the effective diffusion Squire/Lamb constant $\delta \approx 8$ for small scale helicopter rotors.

Figure 6 illustrates the decay of the normalized vortex circulation with the wake age obtained by applying the decay model of Equation (9) and using the value of 2.5 for the decay coefficient. Despite the slope is slightly lower than the measurements reported in Ramasamy et al. [37], a good agreement with the experimental results can be observed.

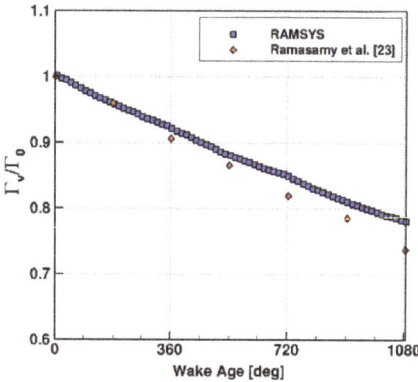

Figure 6. Normalized vortex circulation vs. wake age.

Figure 7 shows the predicted growth of the vortex core radius as a function of the wake age obtained by applying Equation (17) and using Equation (18) for the diffusion parameter δ and Equation (19) for the circulation decay. The picture highlights the close agreement of the calculated parameter δ with the constant value of 8, which is typical for small scale helicopter rotors. The slight increasing deviation with the wake age from the value of 8 is produced by the application of the decay model in the vortex circulation.

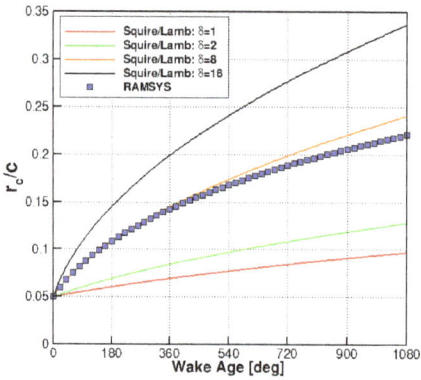

Figure 7. Growth of the vortex core radius as a function of wake age.

4. Results

PIV measurements and BEM numerical simulations were carried out with the main purpose to investigate the structure and development of the rotor blade tip vortices.

4.1. Numerical Test Set-Up and Computational Resources

Each of the four-rotor blades was discretized by 40 chordwise panels and 23 spanwise panels. No rotor hub nor any other body (such as the motor, the controller, etc.) was modelled.

Fourteen rotor revolutions and fourteen wake spirals were required to get a sufficiently long wake and a converged solution. The time discretization adopted corresponded to an azimuth step equal to 2 deg. A computational acceleration was obtained by running the code in parallel mode via the OpenMP API and using 32 cores. The run required about 8 h of elapsed computational time.

4.2. Rotor Wake Ensemble-Averaged Flow Field

The ensemble-averaged velocity field showed the shear layer region surrounding the rotor downwash wake.

The comparison between the experimental results, Figure 8a, and the numerical predictions, Figure 8b highlighted a general similarity but with two main differences:

1. in the experimental PIV the origin of the shear layer is identified at about $x/R = 0.96$ and slightly above $z/R = 0$ because of the deflection produced by the blade elasticity. Instead, the numerical results show the origin of the shear layer exactly at $x/R = 1$ and $z/R = 0$, and this is because the blade was modelled as a fully rigid body;
2. the diffusion produced by the viscous effects causes a marked thickening of the experimental shear layer moving downstream from the rotor disk, while the dissipation produces a reduction of the velocity magnitude which is already visible at around $z/R = -0.7$. These effects are less visible in the numerical results, despite diffusion and dissipation models were applied in the simulations.

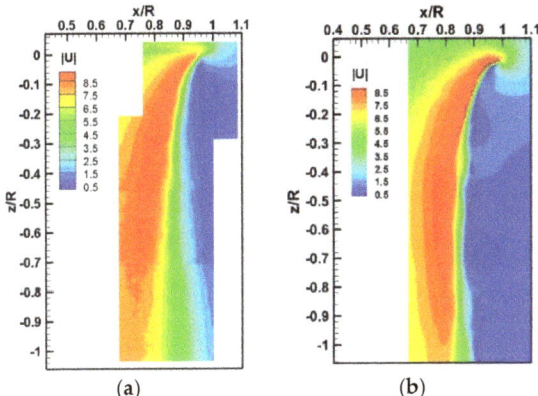

Figure 8. PIV (**a**) and numerical (**b**) ensemble average velocity magnitude colour maps.

A deeper investigation of the aforementioned differences was obtained by comparing the experimental and numerical z component of the induced velocity at several stations below the rotor disk in a region extending up to one radius below the rotor, Figure 9. The numerical results shown in the figure were evaluated at the several azimuthal stations of the last rotor revolution and time-averaged, whereas the velocity fluctuations, due to the flow field unsteadiness, was evaluated and represented in terms of RMS bars. The experimental PIV measurements were made in a fixed vertical plane during about 1450 rotor revolutions. The values were then ensemble-averaged and the velocity fluctuations were represented in terms of RMS bars.

The numerical predictions show a satisfactory agreement with the experiment in the radial region of the blade included from the root cut-out (16%) to the position $r/R = 80\%$, where the maximum of the inflow is measured. In the radial region, where the tip vortex roll-up produces its greater effect, the numerical results show an upwash that is not present in the experiment. Furthermore, discrepancies between the numerical results and the experiment can also be observed in the region of the rotor hub, not modelled in the numerical simulations.

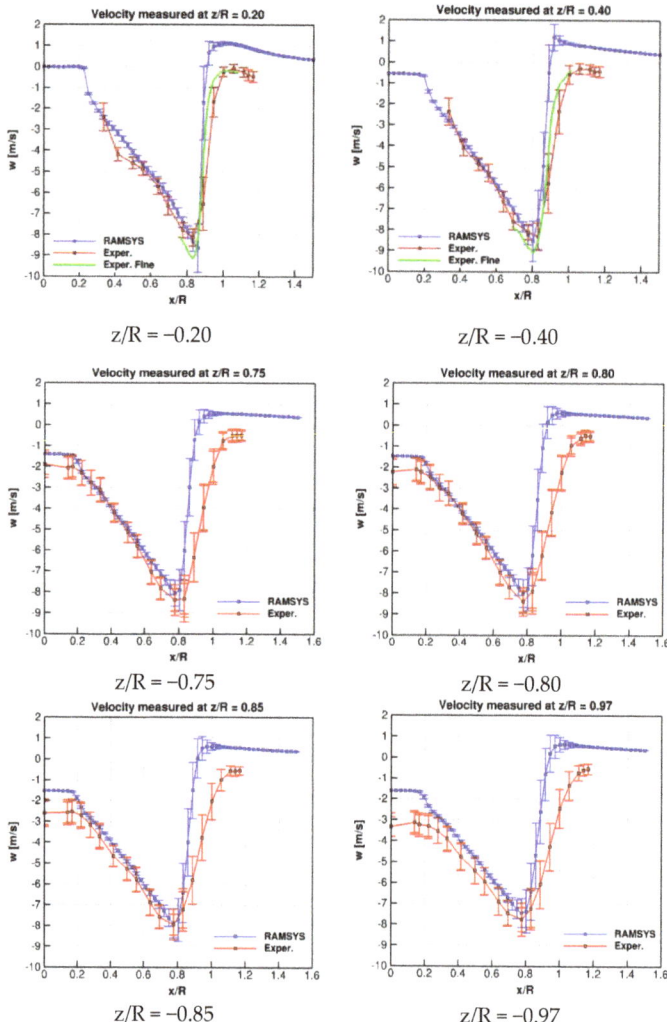

Figure 9. PIV vs. numerical vertical velocity comparison.

Finally, the different slope of the derivative $\partial w/\partial x$ between the experiment and the numerical simulation around 80–100% of the blade radius was further analysed by comparing the PIV measurements made with a finer resolution ($\Delta x = 0.47$ mm), with those at a resolution of ($\Delta x = 2.7$ mm) and with the numerical results evaluated on a grid having a resolution of $\Delta x = \Delta z = 0.6$ mm. The results showed that the slope evaluated by the finer PIV measurements are much closer to the numerical results but this happens up to z/R = 0.4. More downstream, the slope of the finer and coarser PIVs remains almost unchanged.

4.3. Blade Tip Vortices

The application of the Γ_2 method to the PIV and BEM instantaneous velocity fields highlighted the presence of several tip vortices in the measurement regions. Figure 10 shows an instantaneous velocity field in the ROI immediately downstream of the rotor disc. Three tip vortices can be counted and the vortex centres detected by Γ_2 method are shown. Once the centres of the tip vortices were detected, the non-dimensional tangential velocities and circulations were calculated and represented versus the non-dimensional vortex radius (Figure 11).

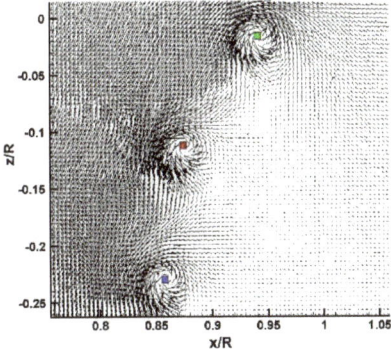

Figure 10. PIV instantaneous velocity fields detected vortex centres.

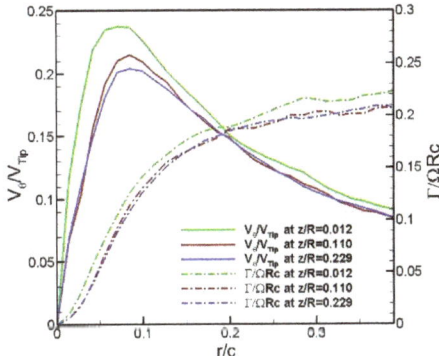

Figure 11. Normalized tangential velocity and circulation vs. r/c for each detected vortex.

The results have shown a decrement of the swirl velocity that follows a second-order trend as the distance from the rotor increases. The circulation decreases as the distance from the disk increases.

The PIV measurements were not phase-locked with the rotor so that a direct comparison between instantaneous velocity fields was not possible. The BEM/PIV comparison was carried out on the tip vortex characteristics versus the distance from the rotor disk. The growth of the normalized vortex core radius with the distance from the rotor disk is shown in Figure 12. The comparison between the experimental and numerical results show that in the latter case the growth is slightly over-estimated.

The distribution of the experimental vortex centres, Figure 13, showed the highest concentration in the proximity to the rotor disk and that their location was enclosed in the shear layer region. Moving downstream, the data scattering increased distributing the centres of the vortical structures both outside and inside the ensemble-averaged downwash, whereas the concentration of vortex decreased due to the fading and/or merging of vortices. The same representation for the numerical vortex centres showed an extremely narrow region of the shear layer and an almost full match between the boundaries of the shear layer region and the path of the vortex centres.

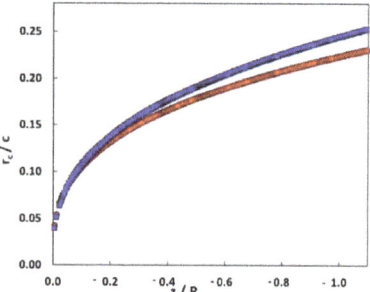

Figure 12. BEM (blue) and Experimental (red) normalised core radius versus the distance from the rotor disk.

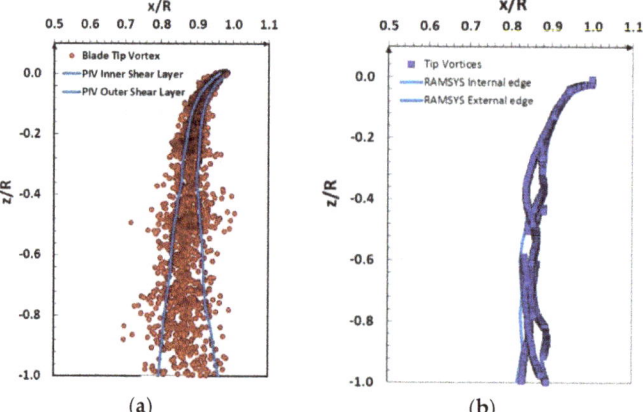

Figure 13. PIV (**a**) and numerical (**b**) position of blade tip vortices with respect to the shear layer region.

Additional characterization of the instantaneous tip vortices was obtained by evaluating the composition of the vorticity orthogonal to the PIV plane $\omega = (\nabla \times V)_\perp$ and comparing the experimental results with the numerical simulations. Figure 14 illustrates the results of this comparison. A cut-off of the vorticity module was set at 1500 1/s to remove as much as possible the flow field small-scale turbulence and to have a more concentrated representation of the tip vortices. PIV results generally show stronger vorticity. The blade flexibility in the experiment generates the tip vortex at a higher and more inboard position with respect to the BEM result ($x/R \approx 0.96$; $z/R \approx 0$ vs. $x/R \approx 1$; $z/R \approx -0.02$). The presence of widespread vorticity along the blade span around $x/R = 0.8$ and $z/R = 0$ can be observed in the PIV due to the blade passage.

The numerical simulations allowed the visualization of the entire wake structure generated by the rotor. In particular, it was possible to visualize the tip vortices and to associate them with the generating blades. Figure 15a shows the numerical tip vortex structure up to a distance of $z = -1.5$ R below the rotor disk. Each of the four colours corresponds to a vortex filament generated by the relative blade. The image illustrates that the vortex structure keeps a geometrical regularity until a distance between $z/R = -0.5$ and -0.6, after which the wake starts showing more chaotic behaviour. A second aspect that is highlighted in the figure is the pairing phenomenon according to which the mutual positions of the vortex filaments tend to interchange after a first rotor revolution. More specifically, looking at the sequence of the tip vortices at about $x/R = +1.0$, it is: cyan-blue-green-red during the first revolution but becomes cyan-green-blue-red during the second revolution. This means that the vortices blue and green begin to roll up with each other until when they have completely interchanged their position after one revolution. Analogously, similar behaviour can be observed at about x/R = −1.0, for the vortex filaments red and cyan. This mechanism contributes to increasing the flow turbulence from the third revolution onward.

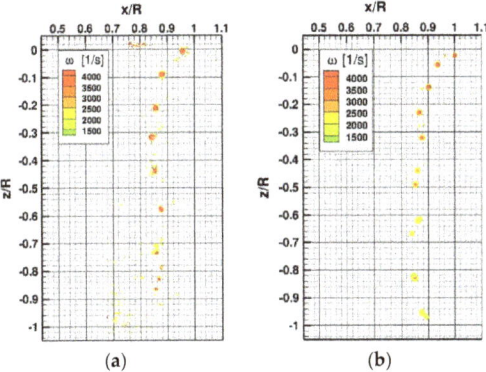

Figure 14. Experimental (**a**) and numerical (**b**) vorticity = $(\nabla \times V)_\perp$ in the PIV plane.

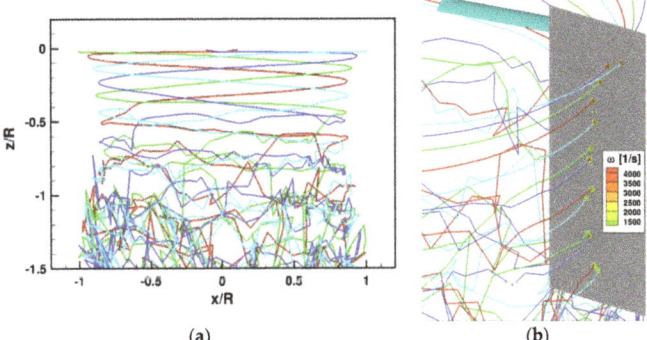

Figure 15. Blade tip trajectories (**a**), Numerical vorticity in the PIV plane and association with the generator vortex filaments (**b**).

Figure 15b shows the association between the numerical vortex filaments and the generated vorticity in the plane corresponding to the PIV measurements. The pairing phenomenon is clearly visible after the second revolution between the green and blue vortex filaments, highlighted in the figure by the black circle, with the vorticity intensity of the first one being smaller than that of the second one. The dashed black circle shows that during the third revolution the green and blue vortex filaments tend to coalesce.

Figure 16 shows a comparison between the experiment and the numerical simulation in terms of the variation of the non-dimensional net circulation of all the identified tip vortices versus the distance from the rotor disk z/R. The net circulation was determined at a distance of 0.25 c from the vortex centre, and by assuming axisymmetric flow in the reference system moving with the vortex core, following the specification in Ramasamy et al. [37]. The experimental data presents three different regions characterised by different slopes, a near-field region comprised between the rotor disk to z/R = −0.23, a mid-region with z/R ranging between −0.23 and −0.75 and a far-field region from z/R = −0.75 to −1.1. In the near-field region, the experimental/numerical comparison shows a reasonable agreement in terms of slope, while an intensity difference of about 17% arises. In the mid-region, where the pairing phenomena start to occur (z/R = −0.5), the experimental slope is almost doubled with respect to the near-field region, and the intensity dissipation is larger than the numerical data. Further downstream, the experimental slope realigns to the numerical trend but with smaller values.

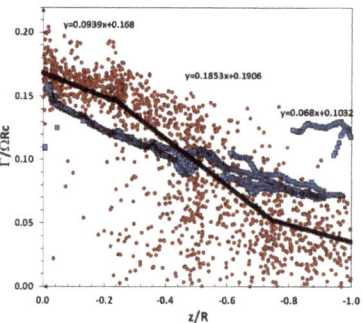

Figure 16. Non-dimensional net circulation. Comparison between the experiment (red) and the numerical simulation (blue). The black lines are the linear interpolation of the experimental data on three separate zones ($0 \geq z/R \geq -0.23$; $-0.75 \geq z/R \geq -0.23$; $-0.75 \geq z/R \geq -1$).

The change in slope of the experimental data in the mid-region and their subsequent scattering at distances $z/R < -0.25$ might be explained by an increase in dissipation due to the growing of turbulence and by the process of pairing and coalescence of vortex filaments, respectively. This process was also mentioned in the explanation of the BEM tip vortices path of Figure 15a. The numerical results resemble the trends illustrated in Figure 6. The merging of co-rotating vortices, which produce a higher circulation, likely causes the presence of higher values in the region $z/R \in [-0.8; -1.0]$.

Finally, Figure 17 shows a comparison between the experiment and the numerical simulation in terms of the maximum values of the non-dimensional swirl velocity of all the identified tip vortices versus the distance z/R. In this case, the experimental vortices also show three distinct zones with a net difference in terms of slopes and intensities. The reason for this behaviour has the same explanation as for the net circulation trend discussed in Figure 16. The numerical results show a typical hyperbolical decay of the velocity intensity according to the model of Equation (16) combined with Equation (17). An interesting match with the experiment can be observed for distances from the rotor disk lower than $z/R = -0.5$. The presence of higher values in the region $z/R \in [-0.8; -1.0]$ is likely caused by the merging of co-rotating vortices which produce a higher velocity swirl, as also mentioned for the net circulation.

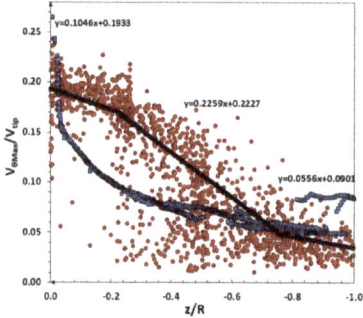

Figure 17. The maximum values of the non-dimensional swirl velocity for each vortex detected. Comparison between the experiment (red) and the numerical simulation (blue). The black lines are the linear interpolation of the experimental data on three separate zones ($0 \geq z/R \geq -0.23$; $-0.75 \geq z/R \geq -0.23$; $-0.75 \geq z/R \geq -1$).

5. Conclusions

Research activity was carried out to characterize experimentally and numerically the blade tip vortices of a small scale four-bladed isolated rotor in hover flight. The investigation of the vortex decay process during the downstream convection of the wake was addressed. 2C-2D PIV measurements were carried out below the rotor disk down to a distance of one rotor radius. The numerical simulations were aimed at assessing the modelling capabilities and the accuracy of a free-wake Boundary Element Methodology.

The Γ_2 vortex centre detection criterion was applied both to the experimental and numerical results. Once detected in the centres, tip vortices were characterised in terms of vorticity, circulation, swirl velocity, core radius and trajectory.

The rotor wake mean velocity field and the instantaneous vortex characteristics were investigated. The experimental/numerical comparisons showed a reasonable agreement in the estimation of the mean velocity inside the rotor wake, whereas the BEM simulations predicted and under-estimated the effect of the diffusion thus generating a smaller shear layer region with respect to the experiment.

The numerical results provide a clear picture of the filament vortex trajectory interested in complex interaction starting at about a distance of $z/R = -0.5$.

The time evolution of the tip vortices was investigated in terms of net circulation and swirl velocity. The PIV results showed similar behaviour for both quantities. They showed a linear mild decay up to the region interested by vortex pairing and coalescence, where a sudden decrease, characterised by a large data scattering, occurred. The numerical modelling predicted a hyperbolic decay of the swirl velocity down to $z/R = -0.4$ followed by an almost constant decay. Conversely, the calculated net circulation showed a gradual decrease throughout the whole wake development.

The comparisons showed discrepancies in the region immediately downstream the rotor disk but significant similarities beyond $z/R = -0.5$.

Author Contributions: Conceptualization, F.D.G.; methodology, F.D.G. and A.V.; software, A.V.; validation, F.D.G. and A.V.; formal analysis, F.D.G. and A.V.; investigation, F.D.G.; resources, F.D.G. and A.V.; data curation, F.D.G., A.V., and G.I.; writing—original draft preparation, F.D.G.; writing—review and editing, F.D.G., A.V., and G.I.; visualization, F.D.G. and A.V.; supervision, F.D.G.; project administration, F.D.G.; funding acquisition, F.D.G. All authors have read and agreed to the published version of the manuscript.

Funding: This research received no external funding.

Institutional Review Board Statement: Not applicable.

Informed Consent Statement: Not applicable.

Conflicts of Interest: The authors declare no conflict of interest.

Abbreviations

The following nomenclature and abbreviations are used in this manuscript:

AR	blade aspect ratio
c	blade chord [m]
CT	thrust coefficient = $T/(\rho \pi R^2 \Omega^2 R^2)$
D	Γ_2 domain radius
K_{tip}	reduced frequency
M_{tip}	tip Mach number
N_b	number of blades
r	local blade radius [m]
r_c	vortex core radius [m]
R	rotor radius [m]
Re_{tip}	tip Reynolds number
t	time [s]
T	rotor thrust [N]

|U|	modulus of the induced velocity [m/s]
v	induced velocity [m/s]
V_B	local tangential velocity = Ω r [m/s]
V_{tip}	velocity at blade tip [m/s]
V_θ	swirl velocity [m/s]
w	axial induced velocity component [m/s]
x	spanwise distance [m]
z	axial distance from the rotor disk [m]
Γ, Γ_v	tip vortex circulation [m^2/s]
Γ_2	vortex detection scalar value
Γ_0	tip vortex circulation at t = 0 [m^2/s]
Γ_2	vortex detection scalar value, [-]
δ	eddy viscosity [m^2/s]
θ_0	blade collective pitch [deg]
ν	kinematic viscosity [m^2/s]
ρ	air density [kg/m^3]
σ	rotor solidity
φ	velocity potential [m^2/s]
Ψ	rotor azimuth angle
ω	out-of-plane vorticity [1/s]
Ω	rotor speed [rad/s]
BEM	Boundary Element Method
CIRA	Centro Italiano Ricerche Aerospaziali
PIV	Particle Image Velocimetry
TE	Trailing-Edge
RMS	Root Mean Square
W	Wake

References

1. Martin, P.B.; Pugliese, G.J.; Leishman, J.G.; Anderson, S.L. Stereo PIV measurement in the wake of a hovering rotor. In Proceedings of the 56th Annual Forum of the American Helicopter Society, Virginia Beach, VA, USA, 2–4 May 2000.
2. Heineck, J.T.; Yamauchi, G.K.; Wadcock, A.J.; Lorenco, L.; Abrego, A. Application of three-component PIV to a hovering rotor wake. In Proceedings of the 56th Annual Forum of the American Helicopter Society, Virginia Beach, VA, USA, 2–4 May 2000.
3. McAlister, K.W. Rotor Wake Development during the First Revolution. *J. Am. Helicopter Soc.* **2004**, *49*, 371–390. [CrossRef]
4. Richard, H.; van der Wall, B.G. Detailed Investigation of Rotor Blade Tip Vortex in Hover Condition by 2C and 3C-PIV. In Proceedings of the 32nd European Rotorcraft Forum, Maastricht, The Netherlands, 12–14 September 2006.
5. De Gregorio, F.; Pengel, K.; Kindler, K. A comprehensive PIV Measurement Campaign on a Fully Equipped Helicopter Model. *Exp. Fluids* **2011**, *53*, 37–49. [CrossRef]
6. Raffel, M.; Bauknecht, A.; Ramasamy, M.; Yamauchi, G.K.; Heineck, J.T.; Jenkins, L.N. Contributions of Particle Image Velocimetry to Helicopter Aerodynamics. *AIAA J.* **2017**, *55*, 2859–2874. [CrossRef] [PubMed]
7. Leishman, J.G. *Principles of Helicopter Aerodynamics*, 2nd ed.; Cambridge University Press: New York, NY, USA, 2006; pp. 771–814.
8. Johnson, W. *Rotorcraft Aeromechanics*, 1st ed.; Cambridge University Press: New York, NY, USA, 2013; pp. 359–365.
9. Antoniadis, A.F.; Drikakis, D.; Zhong, B.; Barakos, G.; Steijl, R.; Biava, M.; Vigevano, L.; Brocklehurst, A.; Boelens, O.; Dietz, M.; et al. Assessment of CFD Methods Against Experimental Flow Measurements for Helicopter Flows. *Aerosp. Sci. Technol.* **2012**, *19*, 86–100. [CrossRef]
10. Duraisamy, K.; Ramasamy, M.; Baeder, J.D.; Leishman, J.G. High-Resolution Computational and Experimental Study of Rotary-Wing Tip Vortex Formation. *AIAA J.* **2007**, *45*, 2593–2602. [CrossRef]
11. Caradonna, F. Performance Measurement and Wake Characteristics of a model rotor in axial flight. *JAHS* **1999**, *44*, 101–108. [CrossRef]
12. Brown, R.; Line, A. Efficient high-resolution wake modeling using the vorticity transport equation. *AIAA J.* **2005**, *43*, 1434–1443. [CrossRef]
13. Mohd, N.N.A.R.; Barakos, G.N. Computational Aerodynamics of Hovering Helicopter Rotors. *J. Mek.* **2018**, *34*, 16–46.
14. Lee, T.E.; Leishman, J.G.; Ramasamy, M. Fluid Dynamics of Interacting Blade Tip Vortices with a Ground Plane. *J. Am. Helicopter Soc.* **2010**, *55*, 22005. [CrossRef]
15. Visingardi, A.; De Gregorio, F.; Schwarz, T.; Schmid, M.; Bakker, R.; Voutsinas, S.; Gallas, Q.; Boisard, R.; Gibertini, G.; Zagaglia, D.; et al. Forces on obstacles in rotor wake—A GARTEUR Action Group. In Proceedings of the 43rd European Rotorcraft Forum, Milan, Italy, 12–15 September 2017.

16. De Gregorio, F.; Visingardi, A.; Nargi, R.E. Investigation of a helicopter model rotor wake interacting with a cylindrical sling load. In Proceedings of the 44th European Rotorcraft Forum, Delft, The Netherlands, 18–21 September 2018.
17. Dallmann, U. Topological Structures of Three-Dimensional Flow Separation. In Proceedings of the AIAA 16th Fluid and Plasma Dynamics Conference, Danvers, MA, USA, 12–14 July 1983.
18. Vollmers, H.; Kreplin, H.-P.; Meier, H.U. Separation and vortical type flow around a prolate spheroid—Evaluation of relevant parameters. In Proceeding of the AGARD Symposium on Aerodynamics of Vortical Type Flows in Three Dimensions, AGARD CP-342, Rotterdam, The Netherlands, 25–28 April 1983; pp. 14.1–14.14.
19. Chong, M.S. A general classification of three-dimensional flow fields. *Phys. Fluids A Fluid Dyn.* **1990**, *2*, 765–777. [CrossRef]
20. Hunt, J.C.R; Wray, A.A.; Moin, P. *Eddies, Stream, and Convergence Zones in Turbulent Flows; Proceedings of the1988 summer program, Report CTR-S88*; Center for Turbulence Research: Stanford, CA, USA, 1988; pp. 193–208.
21. Jeong, J.; Hussain, F. On the identification of a vortex. *J. Fluid Mech.* **1995**, *285*, 69–94. [CrossRef]
22. Graftieaux, L.; Michard, M.; Grosjean, N. Combining PIV, POD and vortex identification algorithms for the study of unsteady turbulent swirling flows. *Meas. Sci. Technol.* **2001**, *12*, 1422–1429. [CrossRef]
23. Mulleners, K.; Raffel, M. The onset of dynamic stall revisited. *Exp. Fluids* **2011**, *52*, 779–793. [CrossRef]
24. De Gregorio, F.; Visingardi, A.; Coletta, M.; Iuso, G. An assessment of vortex detection criteria for 2C-2D PIV Data. *J. Phys. Conf. Ser.* **2020**, *1589*, 012001. [CrossRef]
25. Hart, D.P. Super-resolution PIV by recursive local-correlation. *J. Vis.* **2000**, *3*, 187–194. [CrossRef]
26. Raffel, M.; Willert, C.E.; Wereley, S.T.; Kompenhans, J. *Particle Image Velocimetry—A Practical Guide*, 2nd ed.; Springer: Berlin/Heidelberg, Germany, 2007. [CrossRef]
27. De Gregorio, F.; Visingardi, A. Vortex detection criteria assessment for PIV data in rotorcraft applications. *Exp. Fluids* **2020**, *61*, 179. [CrossRef]
28. Westerweel, J.; Scarano, F. Universal outlier detection for PIV data. *Exp. Fluids* **2005**, *39*, 1096–1100. [CrossRef]
29. Vatistas, G.H. New Model for Intense Self-Similar Vortices. *J. Propuls. Power* **1998**, *14*, 462–469. [CrossRef]
30. Visingardi, A.; D'Alascio, A.; Pagano, A.; Renzoni, P. Validation of CIRA's Rotorcraft Aerodynamic Modelling System with DNW Experimental Data. In Proceedings of the 22nd European Rotorcraft Forum, Brighton, UK, 16–19 September 1996.
31. Morino, L. *A General Theory of Unsteady Compressible Potential Aerodynamics. NASA CR-2464*; National Aeronautics and Space Administration: Washington, DC, USA, 1974.
32. Gennaretti, M.; Bernardini, G. A novel potential-flow boundary integral formulation for helicopter rotors in BVI conditions. In Proceedings of the 11th AIAA/CEAS Aeroacoustics Conference (26th AIAA Aeroacoustics Conference), Monterey, CA, USA, 23–25 May 2005.
33. Scully, M.P. *Computation of Helicopter Rotor Wake Geometry and Its Influence on Rotor Harmonic Airloads*; MIT Aeroelastic and Structures Research Laboratory: Cambridge, MA, USA, 1975.
34. Squire, H.B. The Growth of a Vortex in Turbulent Flow. *Aeronaut. Q.* **1965**, *16*, 302–306. [CrossRef]
35. Bhagwat, M.J.; Leishman, J.G. Generalized Viscous Vortex Model for Application to Free-Vortex Wake and Aeroacoustic Calculations. In Proceedings of the 58th Annual Forum of the American Helicopter Society, Montreal, QC, Canada, 11–13 June 2002.
36. Donaldson, C.; Bilanin, A.J. *Vortex Wakes of Conventional Aircraft*; AGARD: Neuilly sur Seine, France, May 1975.
37. Ramasamy, M.; Leishman, J.G. The interdependence of straining and viscous diffusion effects on vorticity in rotor flow fields. In Proceedings of the 59th Annual Forum of the American Helicopter Society, Phoenix, AZ, USA, 6–8 May 2003.
38. Ramasamy, M.; Leishman, J.G. Reynolds number based blade tip vortex model. In Proceedings of the 61st Annual Forum of the American Helicopter Society, Gravepine, TX, USA, 1–3 June 2005.

Article

Aerodynamic and Structural Strategies for the Rotor Design of a Wind Turbine Scaled Model

Sara Muggiasca, Federico Taruffi, Alessandro Fontanella, Simone Di Carlo and Marco Belloli

Dipartimento di Meccanica, Politecnico di Milano, Via La Masa 1, 20156 Milan, Italy; federico.taruffi@polimi.it (F.T.); alessandro.fontanella@polimi.it (A.F.); simone.dicarlo@polimi.it (S.D.C.); marco.belloli@polimi.it (M.B.)
* Correspondence: sara.muggiasca@polimi.it

Abstract: Experimental tests performed in a wind tunnel or in a natural laboratory represent a fundamental research tool to develop floating wind technologies. In order to obtain reliable results, the wind turbine scale model rotor must be designed so to obtain a fluid-structure interaction comparable to the one experienced by a real machine. This implies an aerodynamic design of the 3D blade geometry but, also, a structural project to match the main aeroelastic issues. For natural laboratory models, due to not controlled test conditions, the wind turbine rotor model must be checked also for extreme winds. The present paper will focus on all the strategies adopted to scale a wind turbine blade presenting two studied cases: the first is a 1:75 scale model for wind tunnel applications and the second a 1:15 model for natural laboratory tests.

Keywords: blade design; wind turbine model; wind tunnel; natural laboratory

1. Introduction

Experiments play a key role in the development of wind turbine technologies. Within the last 20 years, several scale model tests of wind turbines were carried out for very different applications. Wind tunnel tests were run to produce low-uncertainty datasets for the validation of numerical codes [1–3] to test wind turbine control strategies [4–7] and to investigate wakes and wake interaction [8–13]. In floating offshore wind turbine (FOWT), scale-model experiments are required to better understand the fluid-structure interactions between the rotor and the incoming flow. Moreover, numerical tools need data for calibration and verification. Experimental tests can be performed both in indoor laboratories, such as a wave basin and wind tunnel, or in a natural outdoor laboratory. For the first kind of tests, the typical scale allowed is between 1:50–1:100 for wave basin tests and 1:75–1:200 for wind tunnel tests, while higher scales can be adopted for outdoor models. In particular, in the latter case, we can distinguish between prototypal campaigns on close to full-scale wind turbines (see, e.g., [14,15]) and intermediate-scale experiments performed at sea [16,17]. Prototypal activities are generally characterized by a scale lower than 1:10 and a TRL (Technology Readiness Level) of about 7–9 and are generally performed just before commercial development, while intermediate-scale tests are more research-oriented tests, with a TRL of about 5–7 and length scale between 1:10 and 1:20. In any of these cases, to properly reproduce the dynamic behavior of a FOWT, the wave and wind action must be reproduced simultaneously, and this can be obtained by applying different strategies in considerations of the test typology. In wave basin tests, the hydrodynamics is completely reproduced, while the aerodynamic wind turbine behavior can be simulated with different levels of accuracy. The simplest approach uses a solid disc together with a battery of fans to reproduce the drag force; the gyroscopic moments can be obtained through an auxiliary rotating mass [18]. A higher level of accuracy can be ensured with a spinning rotor in the stationary flow [19]. Both these approaches can reproduce just a few issues of the aerodynamic behavior of a wind turbine. A more complete aerodynamic reproduction can be achieved through a more complex setup, where the wind turbine is directly modeled and the wind field is generated by a fan's array [20–22].

Recently, the hybrid/HIL (Hardware In the Loop) approach has been introduced in wave basin tests: the hydrodynamics is physically reproduced during the experiments, while the wind turbine is numerically modeled and connected, in real time, to the physical model [23–27]. The hybrid/HIL approach is adopted in wind tunnel tests on FOWT [28–30]: in this case, the physical model is represented by a scaled wind turbine, while the floating subsystem is emulated by means of a numerical model that is executed in real time. In the case of natural outdoor laboratory tests, wind and waves are naturally present. However, these cannot be controlled. Moreover, the wind and wave statistics are correlated, but their properties and the correlation may differ from the ones expected for a full-scale FOWT. These aspects need to be considered when designing a scale model rotor for outdoor tests, in order to make it representative of the real system. For tests that include a wind turbine physical model, such as hybrid/HIL wind tunnel tests and natural outdoor laboratory tests, the scale model rotor design is of utmost importance. The reduction in size and the limited wind speed results in a Reynolds number that is two/three orders of magnitude lower than in a real rotor. In these conditions, it is hard to match the aerodynamic performance of the full-scale machine with a geometrically scaled copy of its rotor.

This paper presents and compares two scale model rotors of the DTU 10MW [31] reference wind turbine. The first one is a 1:75 scale model for wind tunnel experiments, referred to as the WT (wind tunnel) model throughout the paper, that was designed for hybrid/HIL wind tunnel tests. The second is a 1:15 large-scale model developed within the EU project H2020 Blue Growth Farm, referred to as the BGF (Blue Growth Farm) model. This project aims at developing an offshore multipurpose platform that integrates wave energy converters and a wind turbine with aquaculture. The scaled model of the complete structure will be deployed at the Natural Ocean Engineering Laboratory (NOEL) in Reggio Calabria [32,33]. The key contribution of this paper is to compare two scale models of the same full-scale rotor and to highlight how the specific requirements of a wind tunnel and an outdoor laboratory influence the aerodynamic and structural designs. The outline of the paper is as follows: in Section 2, the scaling strategies for the two models are discussed, in Section 3, the aerodynamic blade design is presented, and in Section 4, the structural design is described. The conclusions are drawn in Section 5.

2. Scaling Strategy

Following the similitude theory, it is possible to represent a full-scale system by means of a scaled model thanks to the definition of dimensionless variables, the so-called scaling factors. In fluid-mechanic systems, such as wind turbines, the problem of scaling results to be over-constrained, mainly due to the interaction among inertia, gravity and viscosity. In particular, for FOWTs, conflicts arise because of the incompatible requirements set by Froude scaling, needed for reproducing the wave and gravity forces, and the necessity of having high Reynolds numbers to be the blade aerodynamic representatives of a full-scale rotor. The definition of the scaling law and the scale factors, computed as in Table 1 as a function of length and velocity, is the consequence of the established target, test modality and campaign goals.

First, the length scale factor is chosen. The main constraints in the selection of the dimension of the model are, on one hand, the test facility capacity and, on the other hand, the costs and complexity related to model-making. In the case of natural laboratory testing, the length scale can be chosen comparing the significant wave height (Hs) at the laboratory site and at the target full-scale deployment site: it is possible to define the ratio between the expected value of Hs at the rated wind speed for the model and the prototype.

Table 1. Factors as a function of the length and velocity for a generic scaling law.

Scale Factor	Symbol	Expression
Length	λ_L	-
Velocity	λ_v	-
Acceleration	λ_a	λ_v^2/λ_L
Frequency	λ_f	λ_v/λ_L
Mass	λ_M	λ_L^3
Inertia	λ_J	λ_L^5
Force	λ_F	$\lambda_L^2 \lambda_v^2$
Power	λ_P	$\lambda_L \lambda_v^3$
Re num. ratio	λ_{Re}	$\lambda_L \lambda_v$
Fr num. ratio	λ_{Fr}	$\lambda_v/\lambda_L^{0.5}$

Second, the velocity scale factor is selected. For a correct reproduction of the aerodynamic effects and, therefore, of the loads acting on the wind turbine rotor, the Reynolds similitude should be satisfied. However, this is rarely possible due to technical constraints: the common length factors to scale a multi-megawatt wind turbine are between 1:10 (outdoor models) and 1:100 (indoor models), and this would imply a model wind speed 10 to 100 times greater than the full-scale one, values out of the operating range for most facilities. Moreover, in the case of FOWT testing, the correct reproduction of the hydrodynamic and gravitational forces becomes important, and this is guaranteed by the Froude similitude. Stating the impossibility of the simultaneous matching of Reynolds and Froude numbers, a compromise has to be reached, according to the type of the test and its purpose.

In the case of natural laboratory testing, another constraint in the velocity selection is given by the site wind occurrence. The met-ocean conditions of the test site must be considered in relation to those expected at the target full-scale deployment site. Differently from the traditional wind tunnel and ocean basin testing, the wind and wave characteristics cannot be controlled; thus, the design of the strategy has to rely on probabilistic data only. Moreover, we also have to consider that the interdependence of the wind and waves cannot be perfectly representative of a target FOWT deployment site. A possible criterion to define the scale velocity takes into consideration the cumulative distribution functions of the wind velocities relative to the test and deployment site: the wind turbine rated hub-height wind speed velocity can be chosen as the value that keeps constant the probability of exceedance, as this would ensure the same probability to have the wind turbine working in partial or full-load conditions for both the model and full scale. Starting with the defined length and velocity factors, the FOWT structure and hydrodynamics are scaled according to Froude's law, while the rotor of the machine is scaled with independent length and velocity scales.

Further details in the wind turbine performance scaling strategies are reported in the following section.

2.1. Rotor Design Requirements

For wind tunnel testing of onshore wind turbines, a possible scaling strategy is to adopt a hybrid scale selecting independently the length and velocity factor and accept a reduction in the Reynolds number at the model scale, coping with it in the aerodynamic design of the model. For the wind tunnel testing of FOWTs, a possibility is to use this approach together with a HIL system, as in reference [28]. For testing a FOWT in a natural laboratory, Froude similitude is mandatory, and the wind speed generation cannot be controlled, as in common facilities; the solution considered here is to adopt a different scaling strategy for the FOWT structure and the rotor. The structure is scaled according to Froude law to ensure the hydrodynamic similitude, whereas the rotor is scaled trying to reproduce the aerodynamic performance of the full-scale turbine [32,33].

The requirements for the aerodynamic design considered here are:

- match the rotor thrust force, as this drives the rigid body motion of the FOWT, the structural loads for the blades and tower,

- reproduce the power as good as possible and
- match the first flapwise bending mode of the rotor.

The matching thrust force is especially important in FOWTs, because it is responsible for the coupled rotor–platform dynamics and the well-known problem of negative damping [34] in the above-rated control.

2.2. The Wind-Tunnel Scale Model (WT)

The wind turbine for wind tunnel tests is a 1/75 scale model of the DTU 10 MW. It was realized as part of the EU H2020 LIFES50+ project to carry out experiments about floating offshore wind turbines. The scale model was designed to: (1) investigate the effects of large platform motions on the rotor aerodynamic loads [35–37] and wake [38,39] and use the experimental data for model calibration/validation and (2) study the global dynamics of FOWT concepts by means of a hybrid/HIL system [28,40]. The scale factors for the wind turbine scale model were defined in consideration of these goals. The length and velocity scale factors were set one independently from the other: (1) because, in hybrid/HIL experiments, it is possible to simulate FOWT dynamics without having to rely on Froude similitude and (2) to increase, as much as possible, the blade Reynolds number (see Figure 1) and have the rotor aerodynamics closer to the full-scale target. The length scale factor was set to 75, the best compromise between having a large rotor while minimizing the blockage effects in the test chamber section of the Politecnico di Milano wind tunnel (14 × 3.84 m). The main constraints for the velocity scale factor are the maximum wind speed it can be achieved in the wind tunnel (15 m/s) and the frequency of the scale model flexible dynamics. The velocity scale factor was fixed to 3. The value of the other scale factors was derived from the mass and velocity scale factors, and these are reported in Table 2.

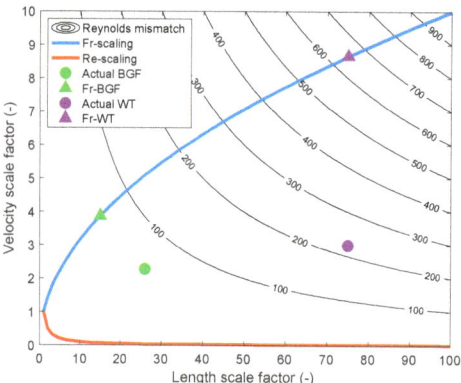

Figure 1. The Reynolds mismatch (reduction with regard to full-scale) as a function of the length and velocity scale factors. The blue and orange lines correspond to the Froude and Reynolds scaling, respectively. Markers show the Blue Growth Farm (BGF) model scaling (green) and wind tunnel model scaling (purple), with "Fr-x" corresponding to the scale factors where Froude scaling was used and "Actual x" corresponding to the adopted scale factors.

2.3. The Blue Growth Farm Outdoor Prototype (BGF)

The wind turbine model designed within the Blue Growth Farm project (EU H2020) is a scaled reproduction of DTU 10 MW, like the previous one. This project has the objective to develop a multipurpose offshore farm placed on a modular floating structure and its targets are efficiency, cost-competitiveness and environmental-friendliness. A wind turbine is combined with wave energy converters and aquaculture to achieve cost-effectiveness in various high-sea applications. This model will be installed at the Natural Ocean Engineering Laboratory (NOEL) in Reggio Calabria.

Table 2. Scale factors for the wind turbine model for wind tunnel testing: Politecnico di Milano.

Scale Factor	Value
Length	75
Velocity	3
Frequency	1/25
Mass	421,875
Force	50,625
Re num. ratio	225
Fr num. ratio	26

The chosen length scale for the global structure (1:15) is appropriate to properly scale the target full-scale deployment site (Golfe de Fos) at the NOEL site. A comparison between the full-scale deployment site and test site wind and waves characteristics considered for the scale definition is shown in Figure 2. The rated wind speed, evaluated imposing the equal probability of exceedance between the test site and full-scale target deployment site, was defined as equal to 5 m/s. The velocity factor and the length factor of the non-Froude scale, used for turbine performances, were calculated accordingly. The resulted scale factors are depicted in Table 3 and Figure 1. This results in having the turbine structure (except the rotor) scaled 1:15 according to Froude law and the performances (rotor dimension included) scaled according to a generic scale defined by the length factor equal to 26 and the velocity factor equal to 2.28. The length factor of 26 was obtained as the factor that, combined with the selected velocity factor of 2.28, gives the same force factor as the 1:15 Froude scale in order to reproduce the 1:15 thrust force. This hybrid scaling approach allows to scale according to Froude law the structure of the platform (crucial for hydrodynamics) and the turbine (i.e., tower dimension and structural frequencies important for a correct representation of the turbine dynamic loads acting on the platform) and, at the same time, to reproduce the Froude-scaled rotor thrust force with a smaller rotor capable of a more accurate reproduction of the aerodynamic loads (lower Reynolds number ratio w.r.t. with regard to Froude) and turbine working conditions occurrence (same probability of exceedance of the rated speed). A comparison between the dimensional values of the DTU 10 MW, WTM and BGF model is shown in Table 4.

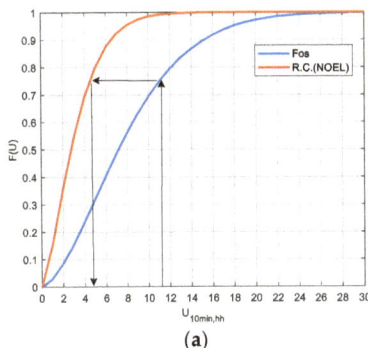
(a)

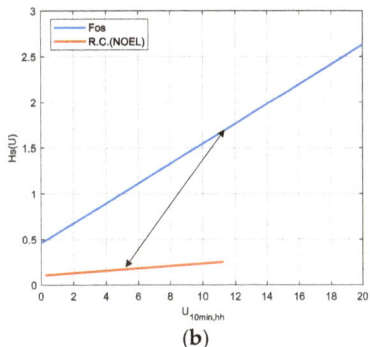
(b)

Figure 2. Comparison between the full-scale deployment site (Fos) and test site at Natural Ocean Engineering Laboratory (NOEL) (R. C.) in terms of the wind speed cumulative distribution (a). Significant wave height Hs as a function of the wind speed (b). $U_{10min,hh}$ is the 10-min mean hub-height wind speed.

Table 3. Scale factors of the hybrid scaling of The Blue Growth Farm outdoor prototype turbine.

Wind Turbine Sub-System	Scale Factor	Value
Structure (Froude)	Length	15
	Frequency	0.258
	Mass	3375
	Inertia	759,375
	Force	3375 [1]
Rotor (non-Froude)	Re num.	58
	Length	26
	Velocity	2.28
	Frequency	0.088
	Inertia	11,881,376
	Force	3514.12 [1]
	Power	8012.19
	Re num. ratio	59
	Fr num. ratio	12

[1] This similitude guarantees that the non-Froude scaled rotor reproduces the Froude scaled thrust force, here identified as the key parameter.

Table 4. Actual gross properties of the DTU 10 MW, wind tunnel model (WTM) and Blue Growth Farm (BGF) model.

Wind Turbine Dimensions	DTU 10 MW	WTM	BGF
Hub height (m)	120	2.1	8
Rotor diameter (m)	178	2.4	6.8
Rated wind speed (m/s)	11.4	3.8	5
Rated rotor speed (rpm)	9.6	240	110
Rated thrust (N)	1.4×10^6	36.7	500
Rated power (W)	10×10^6	78.4	1200

3. Aerodynamic Design

It is possible to divide the procedure for the blade design into two main steps, the first focused on 2D geometry and the second on 3D geometry. The first step is the selection of the airfoil shape, and the second results in the computation of the blade distribution. The aerodynamic design process is similar for the wind tunnel scale model (WT) and the outdoor prototype (BGF).

3.1. Airfoil Selection

The main aerodynamic-related problem arising from scaling is Reynolds number mismatch. The chord Reynolds number for the BGF and WT rotors at their respective rated operating points are shown in Figure 3a. As seen, it is around two orders of magnitude less than for the full-scale wind turbine. To cope with this issue, a new airfoil is selected, and the blade chord and twist distribution is modified. The scale model airfoil is selected according to the operating Reynolds range. Low-thickness airfoils are selected to replace the high-thickness airfoils of the full-scale rotor. The airfoil must have a good lift-to-drag ratio, soft-stall behavior and a linear lift coefficient. One single airfoil or more airfoils can be used along the blade span. Some examples are RG-14, AH79-100C, WM006 and SD7032, SG6040 and SG6041. The airfoils for the BGF and WT models were selected among those of the database "Low Reynolds airfoils" [41]. The low thickness results in a reduced sensitivity to the flow separation at low Reynolds numbers, which translates into a greater lift-to-drag ratio than what would be achieved with a conventional blade profile. The nondimensional shape of the two model airfoils is compared to the one of DTU 10 MW blade (i.e., FFA-W3-xxx) in Figure 4b. It is possible to see that the airfoil selected for the BGF rotor has a higher compared to the WT airfoil. A higher thickness airfoil was selected for the BGF blade, because the expected loads are greater than in a small-scale wind tunnel model, and the Reynolds number is higher thanks to the larger dimensions of the blade.

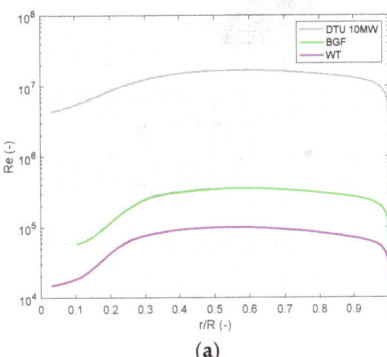
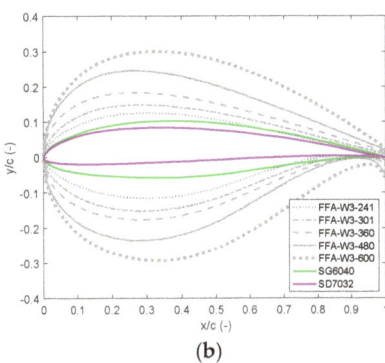

Figure 3. (a) Chord Reynolds number along the blade span at the rated operation for the outdoor prototype (BGF), wind tunnel model (WT) and DTU 10 MW. (b) Nondimensional shape of the airfoil for the outdoor prototype (SG6040), wind tunnel model (SD7032) and DTU 10 MW (FFA-W3-xxx).

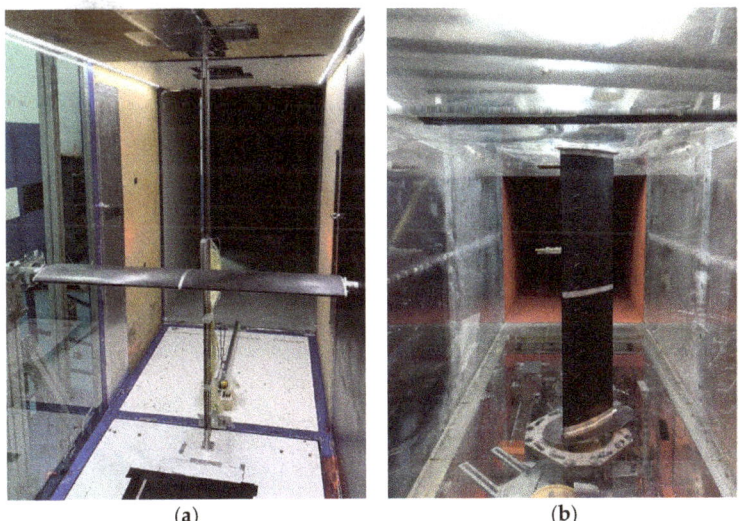

Figure 4. Two-dimensional sectional model tests of the SG6040 airfoil (a) and of the SD7032 (b). In both models, the pressure loop is seen at the midspan and the wake rake downstream of the test section.

The chosen profiles were further characterized through wind tunnel tests on a 2D section model of the airfoil to define the aerodynamic coefficients in a range of Reynolds numbers that is not covered by the literature. Moreover, low-thickness airfoils are sensitive to local separation phenomena and to freestream turbulence. Two-dimensional sectional model tests were carried out with an increased inflow turbulence to simulate the rotor-sampled turbulence seen by the wind turbine model blades; the 2D sectional model was manufactured with similar materials and the production process adopted for the 3D blade, so to have the same surface roughness. The experimental setups for the BGF and WT airfoils are depicted in Figure 4. The lift force was measured with a pressure loop at the model midspan, whereas the drag force was obtained from the wake deficit, which was measured by means of a wake rake positioned downstream of the leading edge. The aerodynamic coefficients of the SG6040 and of the SD7032 that were measured for a Reynolds number of 150,000 are reported in Figure 5. The airfoils behave similarly for small values of the angle-of-attack (AoA), where the lift coefficient is linear and of similar slope. The stall AoA for the SG6040 is slightly lower than for the SD7032, which also shows a larger lift-to-drag ratio. In the BGF blade, aerodynamic performance is traded for structural performance.

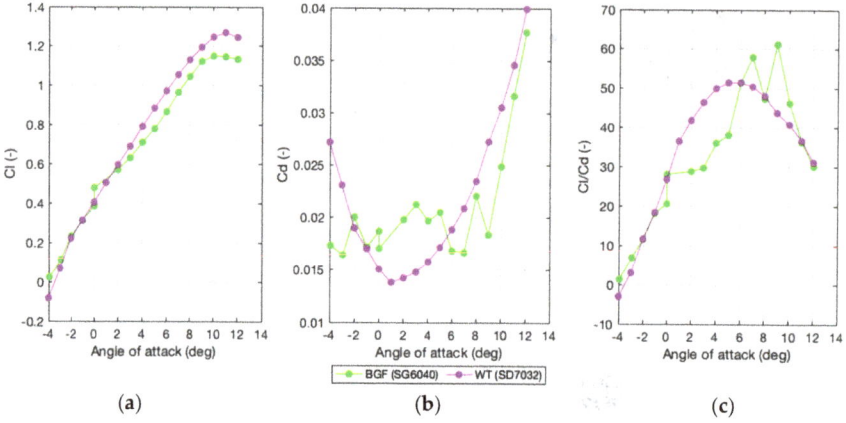

Figure 5. Experimental lift coefficient (**a**), drag coefficient (**b**) and lift-to-drag ratio (**c**) of the blade profiles used in the outdoor prototype (BGF) and wind tunnel model (WT) rotor for a chord Reynolds of 150,000.

3.2. Blade Aerodynamic Design

The goal of the blade aerodynamic design is to preserve the nondimensional thrust force of the full-scale rotor imposing kinematic similarity (i.e., same tip–speed ratio _(TSR)) for the scale model and full-scale blade [3,4,41,42]. The thrust force and the blade–root flapwise bending moment are mainly determined by the lift force, so a good approximation is to match the lift along the blade, section by section. The aerodynamic design procedure was originally introduced in reference [18], but it is recalled here for the sake of clarity. The lift force of any blade section, either of the model or the full scale (FS), is

$$\mathcal{L}_x = \frac{1}{2}\rho v^2 c_x C_{L,x} \tag{1}$$

where subscript x denotes the model or the reference, ρ is the air density, v the airflow speed, c the local chord and C_L the local lift coefficient. The model chord c and twist β are obtained as

$$c_{model} = \frac{c_{RWT}}{\lambda_L} \cdot \frac{K_{L,FS}}{K_{L,model}} \tag{2}$$

$$\beta_{model} = \beta_{RWT} - \frac{C^0_{L,FS}}{K_{L,FS}} + \frac{C^0_{L,model}}{K_{L,model}} \tag{3}$$

where K_L is the slope of the lift coefficient with respect to the angle-of-attack (AoA), C^0_L is the lift at zero AoA and λ_L is the length scale factor for the model. The blade chord is increased, preserving the distribution of the reference rotor, to compensate for the different slopes of the lift coefficients. Given the increased chord, the local twist is modified to achieve the target nondimensional lift force at the rated operating point. A general overview of the algorithm used for the blade aerodynamic design is given in the block diagram of Figure 6. The algorithm takes as input the full-scale chord, twist and thickness distribution, together with the force coefficients of the full-scale airfoils. Based on the theoretical considerations of the thrust matching strategy, the algorithm computes at every iteration the optimal chord and twist distributions to match the thrust force on the rotor. The output of the aerodynamic design of the WT and BGF rotors is shown on the right of Figure 6. The nondimensional chord is similar for the two blades, as the lift coefficients of the SD7032 and SG6040 are comparable. The chord of the BGF blade is slightly larger in the outer portion of the blade, and this is because of the lower lift slope of the SG6040. Similarly, the twist angle for the BGF blade is also lower, because an increased AoA is needed to produce the target nondimensional lift force.

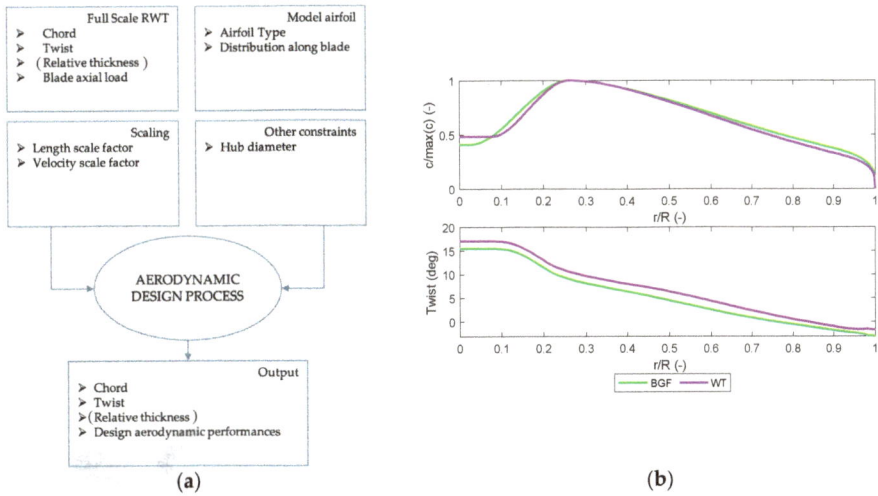

Figure 6. (a) Flow chart of the blade design process (b). Normalized chord and twist distribution of the outdoor prototype (BGF) and the wind tunnel model blade.

The aerodynamic performance of the rotor is evaluated by means of a BEM model implemented in (NREL, Denver, CO, USA), which is one of the most used, freely available rotor aerodynamic solvers. The results are presented in Figure 7 in terms of the power and thrust coefficients for several combinations of TSR and blade pitch angle. The performance of both the scaled rotors is close to the DTU 10 MW. Very small differences are seen in the thrust coefficient, which is largely set by the lift force, objective towards the aerodynamic design. Additionally, the shape is preserved, and this ensures an equal sensitivity to wind, rotor speed and pitch variations. Differences in the power coefficient are slightly more pronounced and are due to the different drag forces developed by the blade profiles. The maximum power coefficient is, in any case, close to 0.5, and it is reached for a TSR around 7.5 and a blade pitch angle of 0 degrees. Additionally, the shape of the power coefficient is similar to the full-scale target, and this enables the implementation in the scale model of conventional power control strategies.

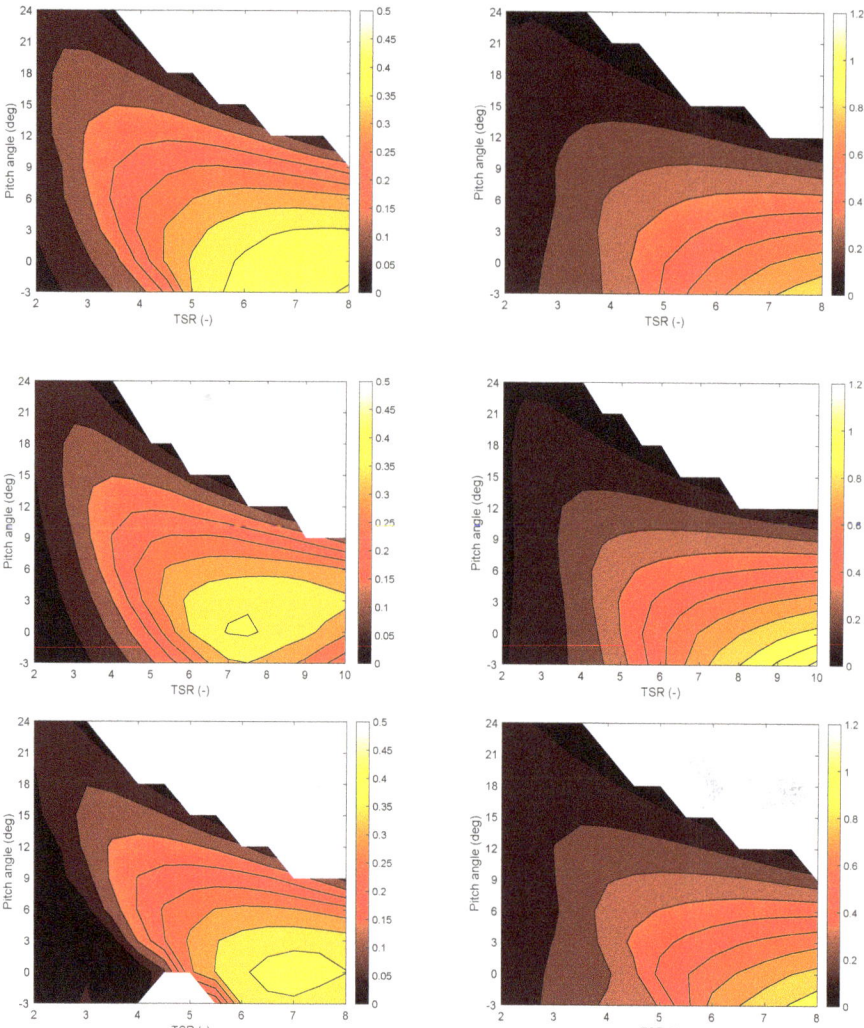

Figure 7. Power coefficient (left) and thrust coefficient (right) of the DTU 10 MW (top), BGF scale model (middle) and WT model (bottom) from the BEM calculations for several values of the tip–speed ratio (TSR) and blade pitch angle. Negative values are omitted.

Comparing the aerodynamic designs of the WT and BGF models, it is possible to conclude that

- the reduced dimension of the rotor requires redesigning the blade to match the full-scale aerodynamic loads. Even in the case of the outdoor prototype, which is three times as big as the WT model, the chord Reynolds is around two orders of magnitude lower than at the full scale.
- low-thickness airfoils have a desirable behavior when using the model Reynolds. In case of the BGF model, the airfoil thickness is increased compared to the WT model to cope with the higher structural requirements that are more stringent. The aerodynamic performance is partially traded for a structural performance.

- the optimization procedure aimed at matching the nondimensional lift force allows to have a model thrust coefficient close to the full scale. The power is reduced, but the shape of the power coefficient is preserved.

4. Structural Design

As described in reference [42], the rotor blades must be designed not only in terms of aerodynamic performances, but, being very flexible components, even the structural parameters must be properly reproduced. For the WT model, structural and aerodynamic design procedures were performed simultaneously, defining the final 3D geometry of the blade as a function of the aerodynamic and of the structural requirements. On the contrary, for BGF model, the structural project was performed once the final 3D geometry was completely defined only based on the aerodynamic requirements. Due to the higher complexity of the structural design of a blade for an outdoor model, it was considered easier to keep the two processes separated.

In both cases the main parameters that must be considered are:

- mass. Mass scales with the cube of the length-scale factor. It is a strict requirement for any scale model rotor. The rotor weight has a significant effect on the flexible dynamics of the wind turbine and the rigid dynamics of the structure, in the case of floating systems. Usually, it is not possible to achieve the scaled mass target, and the blades are designed so to minimize the rotor mass.
- stiffness. Stiffness requirements are set by the need of reproducing the flexible dynamics of the blade. The adoption of low-thickness airfoils makes it difficult to achieve high values of sectional stiffness. Materials that offer a high modulus-to-density ratio, such as CFRP (Carbon Fiber Reinforced Polymer), are utilized for blade manufacturing.

Additionally, for the BGF model, mechanical resistance becomes important. The blade must bear the structural loads, operational and extreme, to which it is exposed.

4.1. Wind Tunnel Scale Model

The output of the aerodynamic design is the chord and twist distribution along the blade span. The structural design aims to define the blade thickness distribution. The blade is divided into three regions, and for each of them, the thickness-over-chord (t/c) is obtained as follows:

- Region 1 is the blade root. The cross-section is circular, and t/c is equal to 1. The radial extension of region 1 is given by manufacturing and assembly constraints. This part of the blade is utilized to fit the components required to mount the blade on the hub.
- Region 3 is the tip region. The cross-section and the t/c are the nominal airfoil selected for the blade design.
- Region 2 is the transition region. The cross-section gradually transitions from a circular shape to the nominal airfoil shape. A longer transition region results in an increased flapwise stiffness, at the expense of a reduced aerodynamic performance.

The radial extension of Region 2 is optimized to have the first flapwise frequency of the scale model blade matching the scaled frequency of the full-scale blade. The radial position where Region 2 starts (i.e., the innermost) is fixed, and the optimization routine searches for the position where Regions 2 ends to minimize the absolute difference between the natural frequency of the first-flapwise mode for the scale model blade and the scaled frequency of the full-scale blade. The optimization routine is the unconstrained nonlinear programming solver fminsearch of MATLAB (The MathWorks, Natick, MA, USA). The natural frequency of the scale model blade is defined based on an Finite Element beam model; the target frequency is obtained from the DTU 10 MW specifications and the frequency-scale factor of Table 2. The radial distribution of the nondimensional thickness for the wind tunnel scale model is shown in Figure 8.

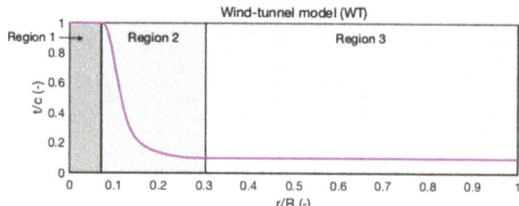

Figure 8. The radial distribution of the nondimensional thickness for the wind tunnel model.

The 3D geometry of the blade is shown in Figure 9. The blade was realized in a composite material starting from an aluminum mold that was manufactured based on the blade external geometry obtained in the design procedure. A single bidirectional CFRP layer was stacked on an inner glass fiber layer. This was necessary in order to increase the torsional stiffness and increased the final mass of the blade of 230 g.

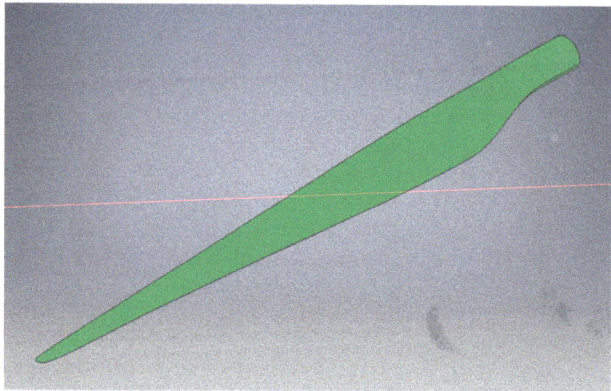

Figure 9. The 3D geometry of the wind tunnel model blade.

The flexible structural dynamics of the blade was assessed via an experimental modal analysis. The frequency of the first flapwise mode is reported in Table 5, where it is compared to the DTU 10MW target, while the mode shapes are displayed in Figure 10. The modes frequencies are lower than expected, and this is mainly because the mass was increased above its design value in the manufacturing process.

Table 5. Comparison between the frequency of the first flapwise mode for the DTU 10 MW (target) and the 1:75 wind tunnel scale model (frequencies are reported at the model scale).

Flapwise Mode	Target Frequency (Hz)	WT Model Frequency (Hz)
First	22.87	17.10
Second	65.25	56.40

4.2. Blue Growth Farm Scale Model

The output of the aerodynamic design for the BGF model is the external shape of the blade. The relative thickness was defined as for the WT model, distinguishing between the three regions presented in the previous section: the aeroelastic design was not the primary goal, and the width of the transition region (Region 2) was defined in order to have the minimum stiffness permitted by the structural design (see Figure 11).

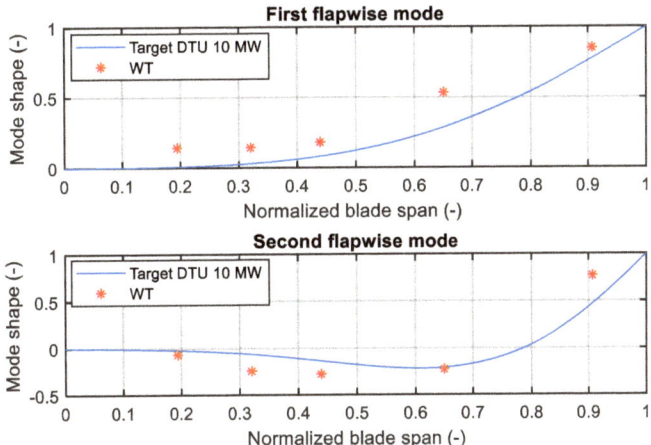

Figure 10. First and second flapwise mode shapes of the wind tunnel scale model blade from the experimental modal identification.

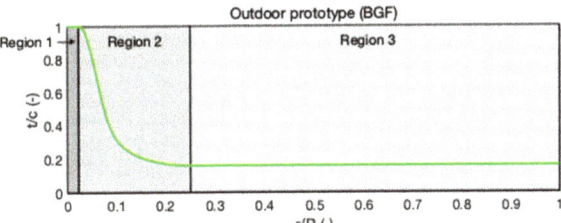

Figure 11. The radial distribution of the nondimensional thickness for the Blue Growth Farm model.

The final 3D external design of the blade is depicted in Figure 12. Two shear webs were inserted along the blade axis to enhance the resistance of the blade to buckling, and they extended from the section at a radius of 0.09 m to the section at a radius of 2.19 m. The blade structure is modeled on the commercial FE software ABAQUS CAE to build a preliminary structural model.

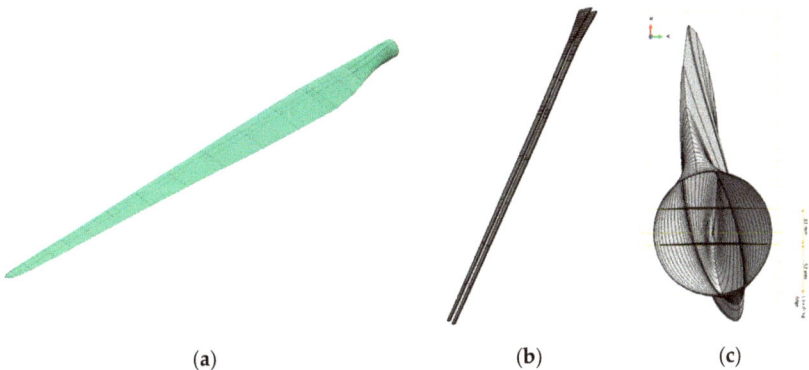

Figure 12. The outdoor prototype blade: external shape (**a**), shear webs (**b**) and their position inside the blade (**c**).

The selected material for the manufacturing of the blade is glass fiber reinforced plastic (GFRP) [43]. In general, fiber-reinforced plastics (FRP) are known for their stiffness, resistance and low weight and are then often adopted for wind energy applications. GFRP is chosen due to its lower cost compared to carbon fiber and for its suitability in marine and warm environments. Two kinds of fiber tissues, bounded by an epoxy resin matrix, were selected to carry out the blade design: S2-glass U (unidirectional fibers) and E-glass Fabric (bidirectional fibers). The blade layup is obtained alternating 0° unidirectional plies and ±45° bidirectional plies. The former is needed to resist the flapwise bending moment, the latter to resist torsional and shear loads. The final layup was defined starting from a preliminary layup that was progressively modified, verifying it with respect to the expected loads. In the first design phase, mechanical resistance was considered as the primary requirement; the design was later refined considering the inertial and aeroelastic requirements. In the second stage, the blade layup was tuned to achieve a match as close as possible to the mass and the scaled natural frequencies. For the structural design, static loads were firstly considered, while dynamic forces due to wind/wave excitation were included only in the final assessments. The static loads were obtained, adopting, as a reference, the IEC 61400-2:2006 [44]. The standard suggests an ensemble of load cases to be withstood by blades, depending on the wind turbine class and the wind condition (normal, occurring continuously during operation and extreme, with a one-year or 50-year period of return). The load cases listed by the standard are many for an industrial-scale wind turbine; however, in the case of a scale model, some of them can be discarded (e.g., fatigue load cases are neglected due to the short duration of the experimental campaign). Among the 10 load cases suggested by [44], the five reported in Table 6 were considered.

Table 6. The evaluated design load cases.

Case Name	U (m/s)	W (rpm)	Collective Pitch (°)	Yaw Angle (°)
Rated	5	101.9	0	0
Rated Yaw	5	101.9	0	30
Park	33	0	90	0
Full Exposure	33	0	0	0
Cut-Out	10.96	109.47	22.67	0

Each load case reported in Table 6 is characterized by four parameters: wind speed, rotor speed, collective pitch and yaw angle. These load cases were simulated in FAST (NREL, Denver, CO, USA), and the loads obtained on each section of the blade, normal and tangential to the rotor plane, were applied to the FE model in ABAQUS to define the stresses and deflections. The distribution of aerodynamic loads per unit length on the blade span for the load cases listed in Table 6 is reported in Figure 13, whereas the resultant blade–root loads are reported in Table 7.

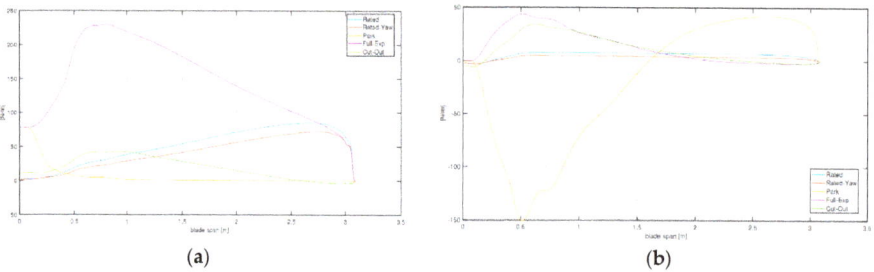

Figure 13. Blade force per unit length: (**a**) normal to the rotor plane (positive downwind) and (**b**) tangential to the rotor plane (positive in the blade rotation direction.

Table 7. Blade–root loads for the design load cases. F_N is the force normal to the rotor plane, F_T the tangential force, M_{FN} the out-of-plane bending moment and M_{FT} the torque.

Case Name	FN (N)	MFN (Nm)	FT (N)	MFT (Nm)
Rated	212.51	495.91	19.40	40.51
Rated Yaw	169.82	402.11	14.37	29.46
Park	31.76	25.62	−85.86	16.18
Full Exposure	608.63	1061.1	56.62	63.74
Cut-Out	82.90	121.64	45.9	62.25

The "full-exposure" case, being the most demanding from the structural point of view, was chosen for dimensioning the layup thickness. The blade–root region carries most of the loads, and the layup thickness is driven by the structural resistance requirements. The tip region does not contribute much to the overall structural resistance; hence, the layup was tuned to match the mass and inertia of the blade. Finally, shear webs have a constant thickness along their span to ease blade manufacturing. The final shell layup thickness is reported in Figure 14.

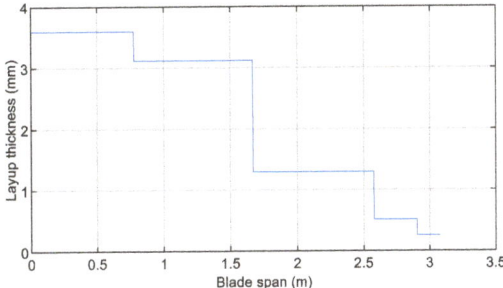

Figure 14. Blade shell layup thickness.

In order to evaluate the structural resistance, the Tsai–Hill criterion [43] was adopted, and the safety factors defined in the GL standards (Germanischer Lloyd standards [45]) were considered. In addition to the static stress analysis, composite laminates can fail by buckling when subjected to compressive or shear loading. Composite structures exhibit large out-of-plane displacements when the initial buckling load is reached, and geometric instability may occur. The linear eigenvalue buckling methodology was chosen among the criteria reported in [43] for the analysis of this phenomenon. The total safety factor for the buckling condition is evaluated from the partial safety factors, as described in reference [45]. The last requirement for the layup design was set on the maximum blade–tip deflection, which must satisfy the minimum tower–blade clearance [46].

The FEM (Finite Element Model) also allows a check on the mass, moment of inertia and natural frequencies of the blade. When designing a scaled blade, it is a common occurrence to obtain natural frequencies that are higher than the ones found by Froude scaling. This happens because aerodynamic constraints force the blade to be shorter than the geometrically scaled one, so that the overall bending stiffness is increased. In natural laboratory models, this fact is, however, not a problem, because a stiffer blade has less deflection, and its frequencies are farther from the dynamic excitation of the rotation (3P). For the BGF model, the mass of the blade is 7.1 kg. This value is lower than the one set by the 1:15 scaling, equal to 12.4 kg, but is higher than the one requested by the 1:26 scaling of the rotor, equal to 2.4 kg. The ideal case would prescribe having a 1:26-scale rotor inertia and a blade structure able to sustain a 1:15-scale force This is impossible to achieve from a resistance point of view. The final blade mass is the best compromise between the resistance and weight of the component, given that the main objective of the whole design is thrust matching. From a dynamic point of view, the final blade design is fairly rigid with respect to the target. This was expected due to the lower length of the blade as compared to the 1:15-scale value, but structural issues were privileged with respect to the aeroelastic ones. The final aeroelastic characteristics of the blades will be described in the following section, as well as the experimental tests performed to verify the blade design obtained from the FE analysis.

Structural Tests

The IEC 61400 [47] prescribes structural tests to assess the dynamic performance of the designed blade and its structural resistance. A complete test campaign would ideally include: (1) measurements of the mass, the mass distribution and the position of the center of mass; (2) check of the natural frequencies and mode shapes; (3) static resistance tests and the (4) ultimate resistance tests, where the blade is brought to failure. For the present application, only the static resistance and modal identification tests were considered mandatory.

For the structural assessment, both fatigue and static tests are requested by the standards. However, given the short duration of the experimental campaign, fatigue tests were omitted. Realistic load conditions were simulated with a static test bench and by means of the load coefficients approach, as proposed in reference [45]. The continuous load distribution along the blade obtained by means of FAST simulations (see Figure 12) was approximated in a discrete number of points by a set of eight concentrated forces, four flapwise and four edgewise. The values of the static loads applied to the blade were computed by means of the shear force-matching method. This method tries to minimize the difference between the area below the internal shear curve with design loads and with the actual loads of the test bench. In this way, it is possible to compute the bending moment generated by the force distribution and verify if the stress state of the tested blade is close to the design one or, at least, more conservative in the regions of interest. A schematic of the mechanism used to apply loads to the blade and the static test bench are shown in Figure 15. The blade is clamped at its root; forces are applied through tie rods that transmit axial forces. Each tie rod is put in a series with a load cell, and the force is applied to the blade section using a saddle. The stress state in the blade is measured with strain gages placed sufficiently far from the tensioning saddles in four different sections, as depicted in Figure 16. The strain gages measure the axial stress due to flapwise and edgewise moments and torsion around the pitch axis. The tip deflection is measured with a laser transducer.

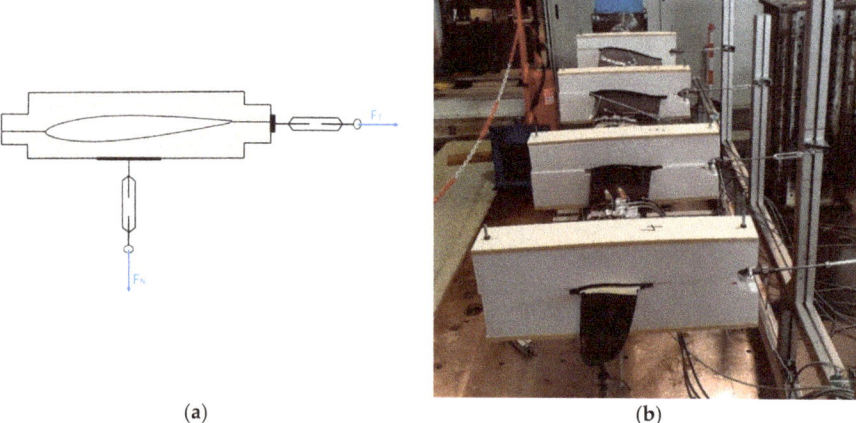

Figure 15. Schematic of the load-application mechanism (**a**). The blade static test bench (**b**).

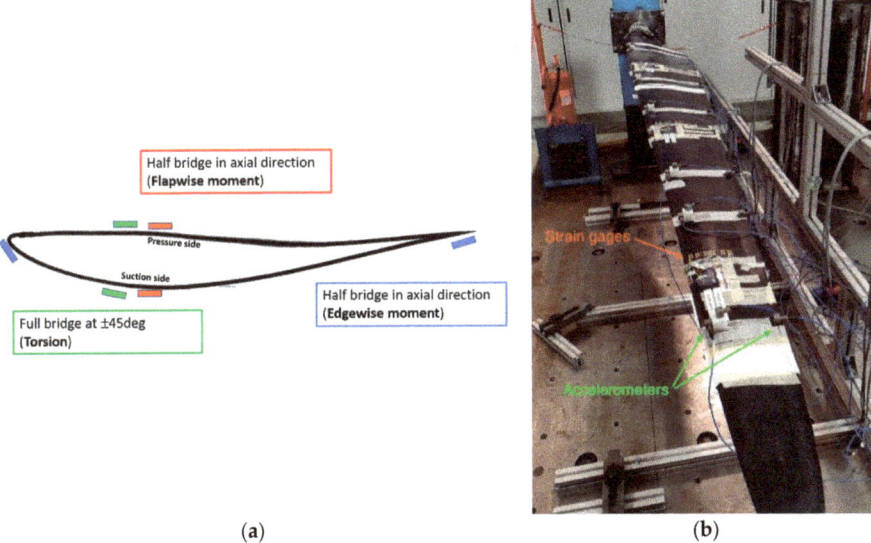

(a)　　　　　　　　　　　　　　　(b)

Figure 16. Strain gages arrangement on a blade section (**a**). Strain gages and accelerometers along the blade (**b**).

Two are the main outcomes of the static test: one side the blade is tested with respect to the design loads, and its resistance is assessed; on the other side, the experimental data can be used to verify the FEM model of the blade. The validated FEM model is then utilized to assess the blade resistance with a more realistic continuous-load distribution. To verify the fidelity of the FEM model, the blade is loaded with a force in the flapwise direction at the tip, and the tip displacement is measured for several load values. A comparison of the experimental data and FEM model predictions is shown in Figure 17.

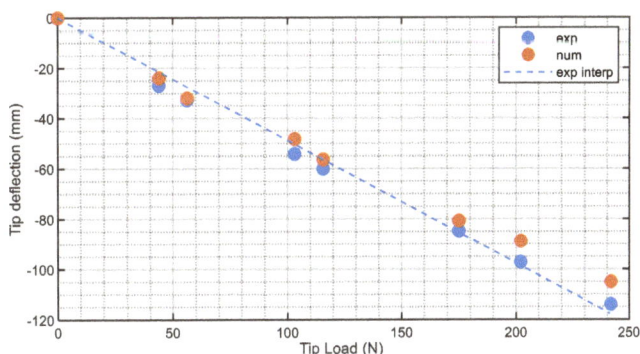

Figure 17. FEA (Finite Element Analysis) and experimental tip deflection.

Natural frequencies and mode shapes are verified with an impact test. The validated FEM model is utilized to define the position of the accelerometers and of the impulse force provided with a dynamometric hammer. Accelerations are measured in six sections of the blade; the flapwise acceleration is measured on both the trailing and the leading edge, while edgewise acceleration is measured only at the leading edge. The comparison between the experimental and FEA results shows a good agreement in terms of the modal shapes, while the frequencies measured are lower with respect to the numerical ones (see Table 8); this discrepancy was imputed to a likely not-perfect modeling of the composite material layup, but the results were considered satisfactory to avoid possible excitation due to the rotation.

Table 8. Comparison between the FEA and experimental flapwise natural frequencies.

Flapwise Mode	FEA (Hz)	Experimental (Hz)
1	9.71	7.82
2	23.54	19.38
3	27.49	23.54
4	54.38	45.11

A comparison between the flapwise modes of the BGF blade and in the DTU 10 MW RWT is reported in Figure 18 and in Table 9. As expected, as the blade shorter with respect to the 1:15 scaling, the natural frequencies of the BGF blade are higher than the target ones; this is not a concern, because aeroelastic properties are considered a secondary target with respect to aerodynamic and structural issues for the natural laboratory blade design.

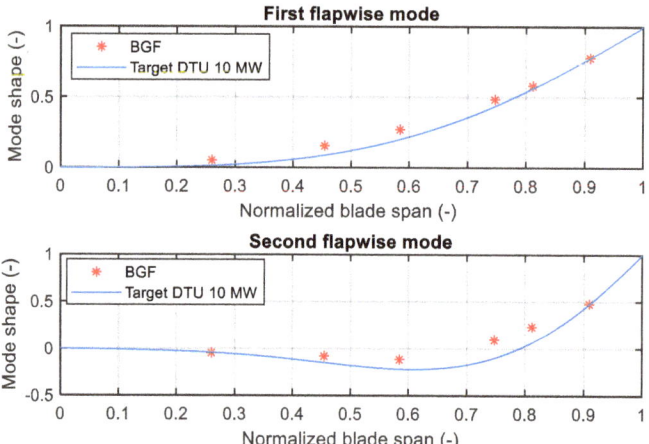

Figure 18. Experimental and target (DTU 10 MW) first and second flapwise modes.

Table 9. Comparison between the frequency of the first flapwise modes for the DTU 10 MW (target) and the experimental frequencies of the outdoor prototype (BGF) (frequencies are reported at the model scale).

Flapwise Mode	Target Frequency (Hz)	BGF Model Frequency (Hz)
First	2.36	7.82
Second	6.74	19.38

Once the design was completed and verified, the blades were manufactured starting from two custom-made molds made of carbon fiber. The dividing line of the molds was chosen in order to follow, as far as possible, the leading edge and trailing edge of the blade. The two molds were manufactured from a master model made on resin using a CNC machine.

5. Conclusions and Recommendations

Scale model experiments about floating wind turbines often rely on Froude-scaled models and a wave basin facility. This paper presents the non-Froude scaled rotor of two 10MW floating wind turbines, with a focus on aerodynamic and structural design. The two models reproduce the DTU 10 MW wind turbine but at different scales and with different purposes: one (1/75) is meant for wind tunnel tests and the other (1/15) for an extended test campaign in a natural laboratory. Both the models were developed for tests that aim to improve the knowledge of wind turbine performances under floating structure excitation and to evaluate the effects of different control strategies.

Important observations were made about the aerodynamic and structural design of a floating wind turbine scale model rotor. It was shown that non-Froude scaling improved the blade Reynolds number and favored the reproduction of the aerodynamic behavior of the full-scale rotor. The Reynolds number for the two models was two-to-three orders of magnitude less than for the full-scale rotor, as a consequence of the reduction in size and the lower wind speed. In the case of wind tunnel tests, these limitations are imposed by the test facility (i.e., maximum flow speed and dimensions of the tests section); in the case of the natural laboratory, wind at the test site is different than the operating wind speed of the full-scale wind turbine. The correct aerodynamic performance is attained through adoption of a performance-scaled design methodology. The same algorithm is applied to the wind tunnel model and the outdoor prototype, and it is shown that by means of non-Froude performance scaling. It is possible to match the thrust coefficient of the reference wind turbine and preserve power coefficient shape. The former is required to simulate at a small scale the thrust force, the blade loads and the wind turbine wake, the latter implementing realistic closed-loop power control strategies. The structural design of the two models is carried out with different goals. In the case of the wind tunnel model, the objective is to reproduce the flexible blade dynamics, and structural requirements are of secondary importance. An optimization algorithm is utilized to define the blade thickness in its innermost region. With this approach, the first flapwise bending mode is matched. The structural design of the outdoor prototype blade is defined by the structural requirements, as the model must operate safely when exposed to uncontrolled environmental conditions. The blade design is guided by the standards for commercial wind turbines; operational loads are obtained from BEM simulations, and the blade material layup is designed with a FEM model in order to withstand these loads.

The following recommendations are proposed as a guide for future model design tasks:

- The aerodynamic design strategy adopted in this article considers just one wind turbine point and modifies the blade chord and twist based on a simple analytical model to match the nondimensional lift. Another possibility is to use a BEM model of the rotor to iteratively simulate several operating conditions and an optimization procedure to define the blade shape that minimizes the difference with respect to a target full-scale performance. The scale model rotor would perform closely to the reference but at the expense of an increase in the design procedure complexity and computational effort.
- The blade design was based on a single airfoil. Additional airfoils, of increased thickness, can be used in the innermost region of the rotor to increase the flapwise stiffness and strength; in this case, this would result in an increase in the design procedure complexity.
- In the case of the outdoor prototype, the aerodynamic and structural designs are achieved by means of two separated analyses, one dictating the blade shape and the other, the material. An improved result might be achieved by means of a more tied analysis. To this purpose, an aeroelastic beam-based model that accounts for the 3D geometry and material data for the blade [48] could be used in place of the 3D FEM model adopted here.

In conclusion, realizing a scale model wind turbine rotor is a challenge, with several objectives and constraints set by the application. The rotor performance is improved when dedicated methodologies and tools are adopted, but the latter are currently lacking. Future research is necessary to develop standardized scale model blade design tools and, therefore, have more reliable scale model experiments and data.

Author Contributions: Conceptualization, F.T. and S.D.C.; methodology, A.F.; formal analysis, S.D.C.; data curation, F.T., A.F. and S.D.C.; writing—original draft preparation, F.T., A.F. and S.D.C.; writing—review and editing, A.F. and S.M.; supervision, S.M. and M.B.; project administration, S.M. and M.B.; funding acquisition, M.B. All authors have read and agreed to the published version of the manuscript.

Funding: This research was funded by Horizon 2020, grant number 774426 and Horizon 2020 grant number 640741.

Institutional Review Board Statement: Not applicable.

Informed Consent Statement: Not applicable.

Data Availability Statement: Not applicable.

Acknowledgments: This paper is partially the result of the work carried out by the authors within the framework of the Blue Growth Farm project, which received funding from the European Union's Horizon 2020 research and innovation programme under grant agreement number 774426, and within the framework of the LIFES50+ project, which received funding from the European Union's Horizon 2020 research and innovation programme under grant agreement number 640741. The content of this work does not report the opinion of the European Commission and reflects only the views of the author(s), including errors or omissions. The European Commission is also not liable for any use that may be made of the information contained herein.

Conflicts of Interest: The authors declare no conflict of interest.

References

1. Hand, M.M.; Simms, D.A.; Fingersh, L.J.; Jager, D.W.; Cotrell, J.R.; Schreck, S.; Larwood, S.M. *Unsteady Aerodynamics Experiment Phase VI: Wind Tunnel Test. Configurations and Available Data Campaigns*; National Tachnical Information Service: Springfield, VA, USA, 2001.
2. Snel, H.; Schepers, J.G.; Montgomerie, B. The MEXICO project (Model Experiments in Controlled Conditions): The database and first results of data processing and interpretation. *J. Phys. Conf. Ser.* **2007**, *75*, 012014. [CrossRef]
3. Berger, F.; Kuhn, M. Experimental investigation of dynamic inflow effects with a scaled wind turbine in a controlled wind tunnel environment. *J. Phys. Conf. Ser.* **2018**, *1037*, 052017. [CrossRef]
4. Bottasso, C.L.; Campagnolo, F.; Petrović, V. Wind tunnel testing of scaled wind turbine models: Beyond aerodynamics. *J. Wind Eng. Ind. Aerodyn.* **2014**, *127*, 11–28. [CrossRef]
5. Azcona, J.; Bouchotrouch, F.; González, M.; Garciandía, J.; Munduate, X.; Kelberlau, F.; Nygaard, A.T. Aerodynamic Thrust Modelling in Wave Tank Tests of Offshore Floating Wind Turbines Using a Ducted Fan. *J. Phys. Conf. Ser.* **2014**, *524*. [CrossRef]
6. Frederik, J.; Kröger, L.; Gülker, G.; van Wingerden, J.-W. Data-driven repetitive control: Wind tunnel experiments under turbulent conditions. *Control Eng. Pract.* **2018**, *80*, 105–115. [CrossRef]
7. Petrović, V.; Berger, F.; Neuhaus, L.; Hölling, M.; Kühn, M. Wind tunnel setup for experimental validation of wind turbine control concepts under tailor-made reproducible wind conditions. *J. Phys. Conf. Ser.* **2019**, *1222*, 012013. [CrossRef]
8. Whale, J.; Papadopoulos, K.; Anderson, C.; Helmis, C.; Skyner, D. A study of the near wake structure of a wind turbine comparing measurements from laboratory and full-scale experiments. *Sol. Energy* **1996**, *56*, 621–633. [CrossRef]
9. Vermeer, L.; Sørensen, J.; Crespo, A. Wind turbine wake aerodynamics. *Prog. Aerosp. Sci.* **2003**, *39*, 467–510. [CrossRef]
10. Wang, J.; Foley, S.; Nanos, E.M.; Yu, T.; Campagnolo, F.; Bottasso, C.L.; Zanotti, A.; Croce, A. Numerical and Experimental Study of Wake Redirection Techniques in a Boundary Layer Wind Tunnel. *J. Phys. Conf. Ser.* **2017**, *854*, 012048. [CrossRef]
11. Campagnolo, F.; Molder, A.; Schreiber, J.; Bottasso, C.L. Comparison of Analytical Wake Models with Wind Tunnel Data. *J. Phys. Conf. Ser.* **2019**, *1256*. [CrossRef]
12. Bromm, M.; Rott, A.; Beck, H.; Vollmer, L.; Steinfeld, G.; Kühn, M. Field investigation on the influence of yaw misalignment on the propagation of wind turbine wakes. *Wind Energy* **2018**, *21*, 1011–1028. [CrossRef]
13. Nanos, E.; Robke, J.; Heckmeier, F.; Jones, K.; Cerny, M.; Iungo, G.V.; Bottasso, C.L. Wake Characterization of a Multipurpose Scaled Wind Turbine Model. In Proceedings of the AIAA Scitech 2019 Forum, San Diego, CA, USA, 7–11 January 2019. [CrossRef]
14. Ruzzo, C.; Muggiasca, S.; Malara, G.; Taruffi, F.; Belloli, M.; Collu, M.; Li, L.; Brizzi, G.; Arena, F. Scaling strategies for multi-purpose floating structures physical modeling: State of art and new perspectives. *Appl. Ocean Res.* **2021**, *108*, 102487. [CrossRef]
15. Viselli, A.M.; Goupee, A.J.; Dagher, H.J. Model Test of a 1:8-Scale Floating Wind Turbine Offshore in the Gulf of Maine1. *J. Offshore Mech. Arct. Eng.* **2015**, *137*, 041901. [CrossRef]
16. Ruzzo, C.; Romolo, A.; Malara, G.; Arena, F.; Taruffi, F.; Muggiasca, S.; Belloli, M.; Bouscasse, B.; Ohana, J.; Santoro, A.; et al. On the arrangement of two experimental activities on a novel multi-purpose floating structure concept. In Proceedings of the Developments in Renewable Energies Offshore, Lisbon, Portugal, 12–15 October 2020; Apple Academic Press: Palm Bay, FL, USA, 2020; pp. 291–302.

17. Ruzzo, C.; Fiamma, V.; Collu, M.; Failla, G.; Nava, V.; Arena, F. On intermediate-scale open-sea experiments on floating offshore structures: Feasibility and application on a spar support for offshore wind turbines. *Mar. Struct.* **2018**, *61*, 220–237. [CrossRef]
18. Cermelli, C.; Roddier, D.; Aubault, A. WindFloat: A Floating Foundation for Offshore Wind Turbines—Part II: Hydrodynamics Analysis. *J. Renew. Sustain. Energy* **2010**, *2*, 135–143.
19. Kraskowski, M.; Zawadzki, K.; Rylke, A. A Method for Computational and Experimental Analysis of the Moored Wind Turbine Seakeeping. In Proceedings of the 18th Australasian Fluid Mechanics Conference, Launceston, Australia, 3–7 December 2012.
20. Chujo, T.; Ishida, S.; Minami, Y.; Nimura, T.; Inoue, S. Model Experiments on the Motion of a SPAR Type Floating Wind Turbine in Wind and Waves. In Proceedings of the ASME 2011 30th International Conference on Ocean, Offshore and Arctic Engineering. Volume 5: Ocean Space Utilization; Ocean Renewable Energy, Rotterdam, The Netherlands, 19–24 June 2011. [CrossRef]
21. Goupee, A.J.; Koo, B.J.; Kimball, R.W.; Lambrakos, K.F.; Dagher, H.J. Experimental Comparison of Three Floating Wind Turbine Concepts. *J. Offshore Mech. Arct. Eng.* **2014**, *136*, 020906. [CrossRef]
22. Shin, H.; Kim, B.; Dam, P.T.; Jung, K. Motion of OC4 5MW Semi-Submersible Offshore Wind Turbine in Irregular Waves. In Proceedings of the ASME 2013 32nd International Conference on Ocean, Offshore and Arctic Engineering. Volume 8: Ocean Renewable Energy, Nantes, France, 9–14 June 2013. [CrossRef]
23. Bachynski, E.E.; Chabaud, V.; Sauder, T. Real-time Hybrid Model Testing of Floating Wind Turbines: Sensitivity to Limited Actuation. *Energy Procedia* **2015**, *80*, 2–12. [CrossRef]
24. Bredmose, A.H.; Campagnolo, F.; Pereira, R.; Sandner, F. Methods for Performing Scale-Tests for Method and Model Validation of Floating Wind Turbines, Deliverable D4.22. Inwind Project. 2014. Available online: https://www.semanticscholar.org/paper/D-4-.-22-%3A-Methods-for-performing-scale-tests-for-Andreas-Manjock/686e6f6e4492dcf71f89ef7620a72ad80bbbf299 (accessed on 26 March 2021).
25. Li, L.; Collu, M.; Gao, Y.; Ruzzo, C.; Arena, F.; Taruffi, F.; Muggiasca, S.; Belloli, M. Development and validation of a coupled numerical model for offshore floating multi-purpose platforms. In Proceedings of the Developments in Renewable Energies Offshore, Lisbon, Portugal, 12–15 October 2020; Apple Academic Press: Palm Bay, FL, USA, 2020; pp. 274–281.
26. Sauder, T.; Chabaud, V.; Thys, M.; Bachynski, E.E.; Sæther, L.O. Real-Time Hybrid Model Testing of a Braceless Semi-Submersible Wind Turbine: Part I-The Hybrid Approach. In Proceedings of the ASME 2016 35th International Conference on Ocean, Offshore and Arctic Engineering, Busan, Korea, 19–24 June 2016.
27. Bachynski, E.E.; Thys, M.; Sauder, T.; Chabaud, V.; Sæther, L.O. Real-Time Hybrid Model Testing of a Braceless Semi-Submersible Wind Turbine: Part II Experimental Results. In Proceedings of the ASME 2016 35th International Conference on Ocean, Offshore and Arctic Engineering, Busan, Korea, 19–24 June 2016.
28. Belloli, M.; Bayati, I.; Facchinetti, A.; Fontanella, A.; Giberti, H.; La Mura, F.; Taruffi, F.; Zasso, A. A hybrid methodology for wind tunnel testing of floating offshore wind turbines. *Ocean Eng.* **2020**, *210*, 107592. [CrossRef]
29. Thys, M.; Fontanella, A.; Taruffi, F.; Belloli, M.; Berthelsen, P.A. Hybrid Model Tests for Floating Offshore Wind Turbines. In Proceedings of the ASME 2019 2nd International Offshore Wind Technical Conference, St. Julian, Malta, 3–6 November 2019.
30. Bayati, I.; Belloli, M.; Bernini, L.; Giberti, H.; Zasso, A. Scale model technology for floating offshore wind turbines. *IET Renew. Power Gener.* **2017**, *11*, 1120–1126. [CrossRef]
31. Bak, C.; Zahle, F.; Bitsche, R.; Kim, T.; Yde, A.; Henriksen, L.C.; Hansen, M.H.; Blasques, J.P.A.; Gaunaa, M.; Natarajan, A. *The DTU 10-MW Reference Wind Turbine*; Technical University of Denmark, DTU Wind Energy: Lyngby, Denmark, 2013.
32. Fontanella, A.; Taruffi, F.; Muggiasca, S.; Belloli, M. Design Methodology for a Floating Offshore Wind Turbine Large-Scale Outdoor Prototype. In Proceedings of the International Conference on Offshore Mechanics and Arctic Engineering, Glasgow, UK, 9–14 June 2019.
33. Muggiasca, S.; Fontanella, A.; Taruffi, F.; Giberti, H.; Facchinetti, A.; Belloli, M.; Bollati, M. Large Aeroelastic Model of a Floating Offshore Wind Turbine: Mechanical and Mechatronics Design. In Proceedings of the ASME 2019 2nd International Offshore Wind Technical Conference, St. Julian, Malta, 3–6 November 2019.
34. Larsen, T.J.; Hanson, T.D. A method to avoid negative damped low frequent tower vibrations for a floating, pitch controlled wind turbine. *J. Phys. Conf. Ser.* **2007**, *75*, 11. [CrossRef]
35. Bayati, I.; Belloli, M.; Bernini, L.; Boldrin, D.; Boorsma, K.; Caboni, M.; Cormier, M.; Mikkelsen, R.; Lutz, T.; Zasso, A. UNAFLOW project: UNsteady Aerodynamics of FLOating Wind turbines. *J. Phys. Conf. Ser.* **2018**, *1037*, 072037. [CrossRef]
36. Bayati, I.; Belloli, M.; Bernini, L.; Zasso, A. A Formulation for the Unsteady Aerodynamics of Floating Wind Turbines, with Focus on the Global System Dynamics. *Offshore Technol.* **2017**, *1*. [CrossRef]
37. Bayati, I.; Belloli, M.; Bernini, L.; Zasso, A. Wind tunnel validation of AeroDyn within LIFES50+ project: Imposed Surge and Pitch tests. *J. Phys. Conf. Ser.* **2016**, *753*, 092001. [CrossRef]
38. Bayati, I.; Belloli, M.; Bernini, L.; Zasso, A. Wind Tunnel Wake Measurements of Floating Offshore Wind Turbines. *Energy Procedia* **2017**, *137*, 214–222. [CrossRef]
39. Bayati, I.; Bernini, L.; Zanotti, A.; Belloli, M.; Zasso, A. Experimental investigation of the unsteady aerodynamics of FOWT through PIV and hot-wire wake measurements. *J. Phys. Conf. Ser.* **2018**, *1037*, 052024. [CrossRef]
40. Bayati, I.; Facchinetti, A.; Fontanella, A.; Taruffi, F.; Belloli, M. Analysis of FOWT dynamics in 2-DOF hybrid HIL wind tunnel experiments. *Ocean Eng.* **2020**, *195*, 106717. [CrossRef]
41. Lyon, C.A.; Broeren, A.P.; Giguère, P.; Gopalarathnam, A.; Selig, M.S. *Summary of Low-Speed Airfoil Data*; SoarTech Pulications: Orlando, FL, USA, 1997.

42. Bayati, I.; Belloli, M.; Bernini, L.; Zasso, A. Aerodynamic design methodology for wind tunnel tests of wind turbine rotors. *J. Wind Eng. Ind. Aerodyn.* **2017**, *167*, 217–227. [CrossRef]
43. Matthews, F.L.; Davies, G.A.O.; Hitchings, D.; Soutis, C. *Finite Element Modelling of Composite Material and Structures*; Woodhead Publishing: Cambridge, UK, 2000.
44. IEC 61400–2. *Wind Turbines-Part 2: Design Requirements for Small Wind Turbines*; International Electrotechnical Commission: Geneva, Switzerland, 2006.
45. Lloyd, G. *Guideline for the Certification of Wind Turbines*; Germanischer Lloyd Industrial Services GmbH: Hamburg, Germany, 2010.
46. Cox, K.; Echtermeyer, A. Structural Design and Analysis of a 10MW Wind Turbine Blade. *Energy Procedia* **2012**, *24*, 194–201. [CrossRef]
47. IEC 61400–23:2002. *Wind Turbine Generator Systems-Part 23: Full Scale Testing of Rotor Blades*; International Electrotechnical Commission: Geneva, Switzerland, 2002.
48. Branner, K.; Blasques, J.; Kim, T.; Fedorov, V.; Berring, P.; Bitsche, R.; Berggreen, C. *Anisotropic Beam Model for Analysis and Design of Passive Controlled Wind Turbine Blades*; DTU Wind Energy: Lyngby, Denmark, 2012.

Article

Power Enhancement of a Vertical Axis Wind Turbine Equipped with an Improved Duct

Mohammad Hassan Ranjbar, Behnam Rafiei, Seyyed Abolfazl Nasrazadani, Kobra Gharali, Madjid Soltani, Armughan Al-Haq and Jatin Nathwani

1. School of Mechanical Engineering, College of Engineering, University of Tehran, Tehran 11155-4563, Iran; mhranjbar@ut.ac.ir (M.H.R.); a.nasrazadani@ut.ac.ir (S.A.N.)
2. Department of Mechanical Engineering, K.N. Toosi University of Technology, Tehran 19395-19919, Iran; behnamrafiei@email.kntu.ac.ir (B.R.); msoltani@uwaterloo.ca (M.S.)
3. Waterloo Institute for Sustainable Energy (WISE), University of Waterloo, Waterloo, ON N2L 3G1, Canada; armughan.al-haq@uwaterloo.ca (A.A.-H.); nathwani@uwaterloo.ca (J.N.)
4. Department of Management Sciences, University of Waterloo, Waterloo, ON N2L 3G1, Canada
5. Department of Civil and Environmental Engineering, University of Waterloo, Waterloo, ON N2L 3G1, Canada
* Correspondence: kgharali@ut.ac.ir

Citation: Ranjbar, M.H.; Rafiei, B.; Nasrazadani, S.A.; Gharali, K.; Soltani, M.; Al-Haq, A.; Nathwani, J. Power Enhancement of a Vertical Axis Wind Turbine Equipped with an Improved Duct. *Energies* 2021, 14, 5780. https://doi.org/10.3390/en14185780

Academic Editor: Alex Zanotti

Received: 12 July 2021
Accepted: 9 September 2021
Published: 14 September 2021

Publisher's Note: MDPI stays neutral with regard to jurisdictional claims in published maps and institutional affiliations.

Copyright: © 2021 by the authors. Licensee MDPI, Basel, Switzerland. This article is an open access article distributed under the terms and conditions of the Creative Commons Attribution (CC BY) license (https://creativecommons.org/licenses/by/4.0/).

Abstract: Efforts to increase the power output of wind turbines include Diffuser Augmented Wind Turbines (DAWT) or a shroud for the rotor of a wind turbine. The selected duct has three main components: a nozzle, a diffuser, and a flange. The combined effect of these components results in enriched upstream velocity for the rotor installed in the throat of the duct. To obtain the maximum velocity in the throat of the duct, the optimum angles of the three parts have been analyzed. A code was developed to allow all the numerical steps including changing the geometries, generating the meshes, and setting up the numerical solver simultaneously. Finally, the optimum geometry of the duct has been established that allows a doubling of the flow velocity. The flow characteristics inside the duct have also been analyzed in detail. An H-Darrieus Vertical Axis Wind Turbine (VAWT) has been simulated inside the optimized duct. The results show that the power coefficient of the DAWT can be enhanced up to 2.9 times. Deep dynamic stall phenomena are captured perfectly. The duct advances the leading-edge vortex generation and delays the vortex separation.

Keywords: DAWT; ducted wind turbine; H type Darrieus; VAWT; dynamic stall; leading edge vortex

1. Introduction

Boosting the annual energy extraction of wind turbines has been an important focus of wind energy researchers. Increasing the swept area of a wind turbine is a well-established study option. Since 1951, when the first wind turbine was connected to the grid, attempts to increase the size of rotors have increased [1]. Another way to extract more wind energy is the insertion of a rotor of a wind turbine in a duct. Ducts can also be applied to airborne and windmills [2,3]. A wind turbine located within a duct is referred to as the Diffuser Augmented Wind Turbine (DAWT). It is also known as a ducted or a shrouded wind turbine. This turbine is installed within the circular section that causes an increase in the mass flow to the rotor due to its sectional circulation. Thus, the power extracted by the turbine increases, and the Betz limit can be surpassed both in terms of the rotor and duct-exit area [4]. The additional advantages of DAWTs are the decreased cut-in speed, the minimized tip losses [5], and noise [6]. Additionally, there are the benefits of low sensitivity to variations of the yaw angle [7], and the prospects for use in airborne applications for harnessing the stable and strong high-altitude wind streams [3,8], as well as installations in an urban environment [9,10]. Several investigations have been conducted experimentally and numerically for evaluation of the performance, the torque, and the thrust of DAWTs [11,12]. Moreover, DAWTs with auxiliary slots, high-lift compact DAWTs, conical DAWTs, and ground-based DAWTs have been considered in previous studies [13,14]. Ohya and Karasudani proved experimentally that inserting a horizontal axis wind turbine within a duct can increase the extracted power up to four times for some wind speeds [15].

Heikal et al. [16] studied the power coefficient (C_P) of a ducted wind turbine with numerical simulations. They reported a power factor of 90% for some wind speeds. Wind turbine ducts have different types and parts. For studying the wind turbine power increase due to ducts, it is necessary to examine various components of ducts. A duct may contain a nozzle, a diffuser, and a flange. The combined effects of these components on performance are not clear and only a few aspects have been investigated. In the study by Rochman et al. [17], only the geometry of the flange of the duct was examined, which showed that adding a flat flange at the end of the duct causes an increase of 29% of wind speed in the throat region of the duct in comparison to the case without a flange. The studies of Ranjbar et al. [18] and Al-Zahabi et al. [19] focused only on the angle of a flange. Al-Zahabi et al. showed that, for their selected geometry, by setting the angle of the flange at 15 degrees, the power coefficient can be increased up to 5%. Ranjbar et al. demonstrated that if the duct is connected to a nozzle, the throat velocity can be enhanced up to 5.4%. The most prominent study of the effects of the angle and the length of diffusers can be found in the work of Ohya and Karasudani [15]. They tested four ducts with different diffuser lengths. They showed experimentally that increasing the length of the diffuser makes reaching high wind speeds in the throat region of the duct possible. Kosasih and Tondelli [20] studied both the length of the flange and the length of the diffuser. In the examined cases, they proved that the flange length increment made an increase in the wind velocity of the throat. On the contrary, the increased length of the diffuser resulted in a reduction of the ultimate power coefficient of the turbine located in the duct.

Wang et al. experimentally investigated the aerodynamics performance of a Vertical Axis Wind Turbine (VAWT) inside a diffuser. The diffuser increased the maximum generated power up to 26.31% [21]. Watanabe et al. [22] had experimental investigations of a ducted VAWT. Their optimization was based on changing the lengths and angles of the diffuser and flange. They also performed a parametric study on chord length, blade thickness, solidity, and Reynolds number. As they reported, the power extraction was doubled. Ghazalla et al. [23] searched a proper location for the rotor inside the duct. The best position was the throat of the duct with the smallest cross section. Zanforlin et al. [24] studied bi-directional symmetrical diffuser VAWTs at different yaw angles. The power coefficients of bi-directional symmetrical diffuser VAWTs were higher than a bare turbine at any yaw angle. Hashem and Mohamed [25] used a Darrieus VAWT with three different types of the duct. They compared a flat-panel duct, a curved-surface duct, and a cycloidal-surface duct and showed the rotor covered with the cycloidal-surface duct generates the most power. Amgad et al. [26] investigated the aeroacoustics and aerodynamics of a ducted VAWT. The results indicated that a velocity increase in the throat of the duct can augment the power coefficient by about 82%, which results in noisier operation. The urban integration of diffuser VAWTs was presented by Zanforlin and Letizia. They installed VAWTs inside a roof-and-diffuser system and showed that the power can be enhanced by about 40% rather than a case without a diffuser [27].

One of the complicated phenomena for blade studies is dynamic stall. The sudden increase of the aerodynamic loads results in problems for the performance of wind turbines [28–31]. Although for horizontal axis wind turbines, the dynamic stall can be controlled, for vertical axis ones, these phenomena always exit [32,33] and should be considered.

The angles of the nozzle, diffuser, and flange as the main components of the duct play an important role to increase the wind speed inside the duct. Here, the angle optimization of these three components of the duct was considered, which had not been studied before to the best of the authors' knowledge. The optimum angles will be achieved when the velocity inside the throat of the duct reaches its maximum value. A code has been developed to perform numerical simulations automatically. Then an H-Darrieus rotor was simulated inside the optimum duct. The power coefficient increment and the flow structure were investigated in detail. For the previous studies, the dynamic stall phenomena were investigated for ducted vertical wind turbines in detail, which were considered here.

2. Geometry and Numerical Approach of Duct Study

For the governing equations of the numerical approach, the continuity equation derived from the conservation of mass law and the total net flux of the control volume in the tensor form is

$$\frac{\partial \rho}{\partial t} + \frac{\partial v_i}{\partial x_i} = 0, \tag{1}$$

while the time-dependent term of the equation, $\partial \rho / \partial t$, is zero. To describe the fluid interaction, Navier–Stokes equations were used. The tensor form of these equations is given by

$$\rho \frac{\partial v_i}{\partial t} + \rho \frac{\partial (v_i v_j)}{\partial x_j} = -\frac{\partial P}{\partial x_i} + \mu \left(\frac{\partial^2 v_i}{\partial x_j^2} + \frac{\partial^2 v_j}{\partial x_j \partial x_i} \right) + G_i, \tag{2}$$

where v, ρ, P, and μ present the velocity, fluid density, static pressure, and dynamic viscosity, respectively. In Equation (2), G_i, body force, is ignored because of its negligible value. The flow is considered two-dimensional and incompressible.

For solving full-elliptic Navier–Stokes equations, k-ε and k-ω models were used. According to the previous studies [18,19,34], these two models can provide reliable results for flows around a rotor located in a duct. Two models were tested, and the k-ω model was selected [35].

The finite volume method was applied. The governing equations were calculated via SIMPLE for the optimization part. For the simulation of the wind turbine with and without the duct, PISO was utilized. The discretization of equations was based on the second-order upwind method. The convergence criterion of the numerical simulations was 10^{-3} for all equations. This threshold was absolute and a residual normalization was done.

Various types of ducts were designed and built for wind turbines. The duct used in this study was the duct provided by Ohya et al. [15,36]. This duct consisted of three parts: a nozzle, a diffuser, and a flange. The aim of using the duct was to increase the extracted power of the wind. The duct increases the kinetic energy of the flow by raising the wind speed in the throat region [37]. Each part of the duct plays an effective role in increasing the flow velocity in the throat. The nozzle increases the cross section of the input, the diffuser enhances the flow velocity profile in the throat, and the flange creates a low-pressure region by creating strong vortices downstream of the duct. The downstream, low-pressure area creates a pressure difference between upstream and downstream of the duct. The pressure difference results in the flow velocity and the kinetic energy increments in the throat, where the turbine is installed [38]. Considering the mentioned geometry shown in Figure 1, the effective factors are the angles of these three parts. In Figure 1, the nozzle (α), the diffuser (β), and the flange (γ) are shown. In the current study, to determine an optimum angle for each part, the lengths of the nozzle, diffuser, and flange were assumed constant with the values of 0.3D, 1.3D, and 0.3D, respectively.

Since the two-dimensional section of the duct is symmetrical, instead of simulating the entire domain, only half of it was considered as the numerical domain. Decreasing the size of the domain reduces computational costs. Although this simplification has been used and verified by other researchers [39], the authors also solved the complete domain and compared the results. The top wall had the slip wall condition, which reduced the influences of the shearing stress [39]. For the boundary conditions of the inlet and outlet of the domain, the velocity inlet and the pressure outlet were considered, respectively. The inlet velocity was equal to 8 m/s for all duct simulations. This velocity was considered as the reference velocity for making the flow parameters dimensionless. The wall and the inlet and the outlet of the domain were located far enough from the duct to reduce the effects of the boundaries on the duct.

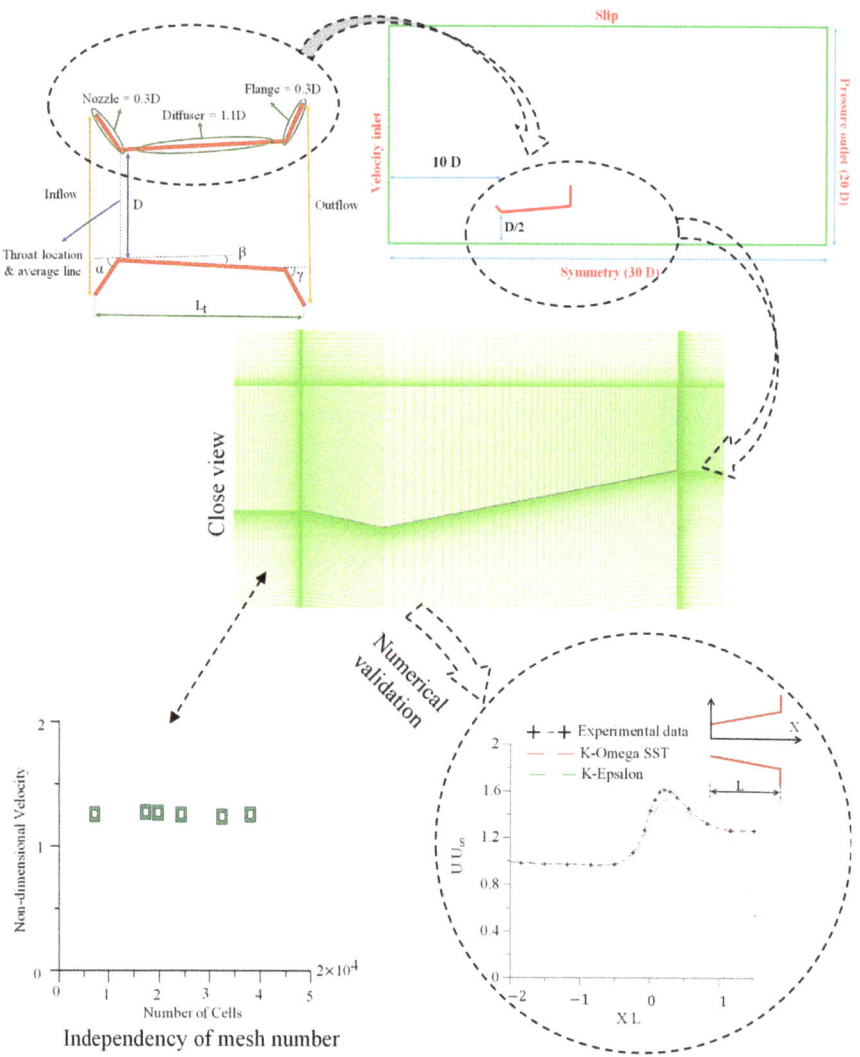

Figure 1. Numerical approach for duct study and validation by experimental results of Ohya and Karasudani [15].

Quadratic meshes were generated inside the computational domain (Figure 1). The growth rate of the mesh length in the walls was 1.8. The maximum y+ values were less than 3 except in some limited cells close to the walls, which were about 8. For getting the optimum geometry of each component of the duct, all parts were needed to mesh separately in their blocks. Mesh independence was important for providing proper answers. Since the optimum angles of the duct were not accessible before optimizing the duct, the angle values of the nozzle, diffuser, and flange were considered as 11, 12, and 90 degrees, respectively, based on the study of Ohya et al. [15,36]. In Figure 1, the number of cells examined for the duct is from 14,400 to 76,000 cells. The average flow velocity in the throat was dimensionalized with the reference velocity (8 m/s). From the results, meshes with 70,000 cells were independent. Increasing the number of cells caused around 4.14% velocity deviation. Considering the computational cost, meshes with 70,000 cells were selected.

For ensuring a reliable numerical simulation, the results were validated against the experimental data provided by Ohya et al. [15,36]. The experimental setup had no nozzle, shown in Figure 1, and the same structure was applied for the numerical simulation. The non-dimensional velocities were measured in the central line of the duct. X is the centerline axis of the duct. L is the length of the diffuser. The diffuser starts at X = 0. The flange is located at X = L with an angle of 90 degrees. Two turbulence models k-ω and k-ε were applied. The comparison indicated that the k-ω model had less error than the k-ε model. For the k-ω model, the maximum discrepancy of the results compared to the experimental results was around 5%. Therefore, the k-ω turbulence model was used for the rest of this study.

3. Optimization

Since the optimum angles of the three parts including nozzle, diffuser, and flange were studied, a code was developed (described in Figure 2). Based on this code, the optimum duct was the duct that had the highest velocity in the throat.

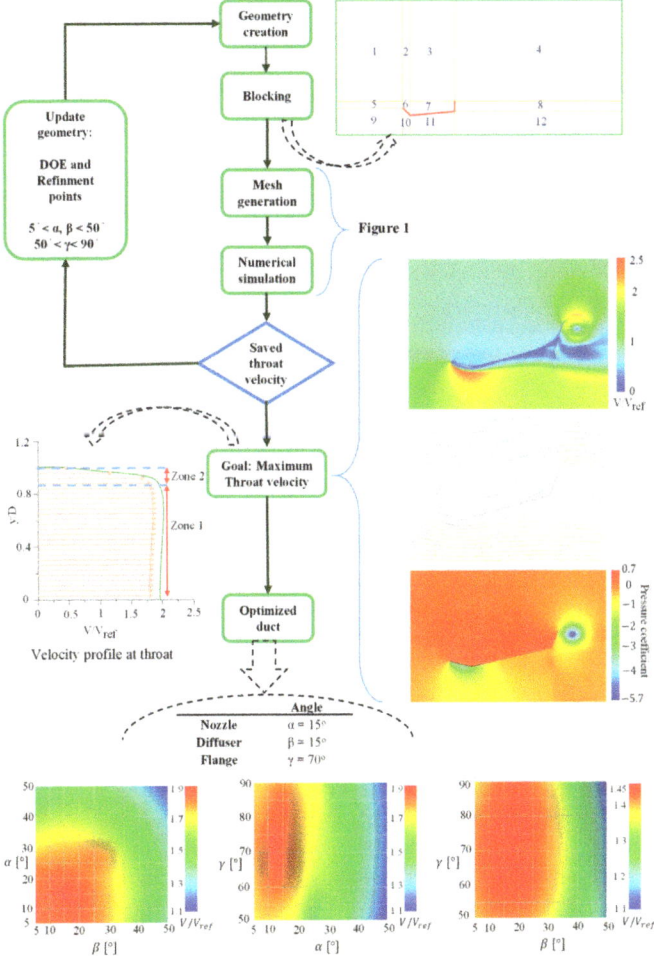

Figure 2. Optimized duct algorithm and final duct geometry.

3.1. Geometry Creation

The angles of the nozzle and diffuser were varied between 5–50 degrees and the angle of the flange was varied between 50–90 degrees. In the first step of the code, the duct geometry must be created. The Design of Experiment (DOE) method was used to reduce the computational cost and increase the accuracy. A common method of DOE is the Central Composite Design (CCD) method presented by Box and Wilson [40]. The number of samples in this method is $2^k + 2k + 1$, where k is the number of input variables [41]. For three input variables including angles of the nozzle, diffuser, and flange, 15 samples are assumed.

3.2. Blocking

For the numerical simulation codes with variable geometry, generating an unstructured mesh is more convenient. One of the strengths of the current code is creating an automatic structured mesh in each loop while the geometry varies. The most challenging part was the creation of blocks. As shown in Figure 2, the domain of the simulation was divided into 12 different blocks. In each loop, the blocks were modified based on the new geometry.

3.3. Structured Grid

With the aid of the blocks, a structured mesh was generated automatically in each loop. Then the numerical solver can use the advantages of the structured mesh and provide appropriate results. More details of the mesh can be found in the mesh generation of the duct.

3.4. Numerical Simulation

In this step, the numerical setup explained in the numerical approach section was applied to the constructed meshes. During iteration, the resultant average velocity of the throat was stored in a file and transferred to the post-processing step.

3.5. Kriging Method

With the help of surrogate methods, the DOE data can be interpolated. One of the simplest and lowly accurate methods of interpolation is a second-order polynomial method. A more accurate method is Kriging or Gaussian process regression, which is a stochastic interpolation method. The Kriging method was used for the current study since it was applied for the optimization of wind turbine-related studies successfully. Dimitrov et al. [42] applied the Kriging method to estimate wind turbine site-specific load. Kumar et al. [43] optimized blade thickness of horizontal axis tidal stream turbine by Kriging model. Solt et al. [44] used the Kriging surrogate model and polynomial chaos expansion for a wind turbine.

In order to increase the accuracy of the Kriging model, a number of refinement points were added in the empty design spaces. Depending on the number of input variables, the distribution of the samples in the design space, and the amount of nonlinearity of the problem, the number of refinement points varies. Here, 16 refinement points were added to improve the accuracy of the results.

3.6. Maximum Throat Velocity

Because of the continuity, the maximum average speed should be in the smallest area of the duct, called the throat. Usually, the turbine is located in the throat [15]. The velocity profile of the optimum duct in the throat is demonstrated in Figure 2 (left). The velocities in the throat were averaged along the average line. The velocity distribution in zone 1 was almost uniform with the velocity deviation of 2.4%; this amount can be negligible. Close to the wall of the duct, because of the no-slip condition, the velocity suddenly decreased (Zone 2). The boundary of zone 2 was where the velocity reaches 99% of the average throat velocity. It can be concluded that the maximum kinetic energy was in zone 1, which covered about 90% of the throat area. Considering the rotor installation in the throat, a clearance between the tip of the blades and the duct was required [5]; then, the blades experienced almost uniform upstream velocity.

Ohya and Karasudani [15] compared the entrance flow into a nozzle and a diffuser. From their visualizations, the upstream streamlines converged into the entrance of the diffuser while the upstream streamlines diverged into the entrance of the nozzle. They concluded that because of this behavior of the flow, the velocity in the throat of the diffuser was higher than that of the nozzle. Streamlines, velocity, and pressure contours (Figure 2, right) showed that the nozzle and the diffuser caused velocity intensification in the throat. The combination of both the nozzle and diffuser resulted in converging the streamlines before entering the duct. Then, this combination increased the flow velocity in the throat of the duct. The reduced cross section in the throat with a constant mass flow rate resulted in denser streamlines. In this figure, the velocity of the flow in the throat is almost doubled. Additionally, the velocity of the flow decreased after passing the throat. The contours of the pressure proved the pressure reduction through the duct and in the wake, revealing the tendency of the flow to pass through the duct. The flange with an almost vertical angle to the wind direction caused the vortex shedding behind the duct. The wake vortices resulted in the flow velocity oscillation inside the throat. Because of these oscillations, the average dimensionless velocity in the throat was considered.

3.7. Optimized Duct

In the last step of the code, a duct can be chosen as the optimum duct when the highest velocity has occurred in the throat. The angles of the optimum duct are shown as the last step in Figure 2. The dimensionless lengths of 0.3D, 1.1D, and 0.3D with the angles of 15, 15, and 70 degrees for the nozzle, diffuser, and flange, respectively, provided the maximum average velocity in the throat, while the average velocity ratio of the throat was 1.97.

4. Ducted Wind Turbine: Numerical Approach

Experimental and numerical investigations on bare wind turbines have shown that their maximum power coefficient is in the range of Betz-limit [45,46]. In order to investigate the effects of the duct on the performance of wind turbines, two simulations of a VAWT with and without the duct were done. The bare turbine was an H-Darrieus wind turbine, according to Bravo et al. [47] (Figure 3). The blade angle was an angle between the chord of the blade and the radius of the rotor. Since the chord of each airfoil was always tangential to the rotor, the blade angle was 90 degrees. The duct with optimum geometry presented in Figure 2 was considered. Since the H-Darrieus wind turbine inside the duct had a diameter of 250 cm, the lengths of the nozzle, the diffuser, and the flange were considered as 75 cm, 275 cm, and 75 cm, respectively. The length of the throat was 277.5 cm. To insert the rotor inside the duct, the clearance between the rotor and the duct was considered. In ducted turbines, the gap between the duct and the tip of the blade has an important role. Since there is a low-pressure region inside the diffuser, there is a possibility of stall. This gap creates a highly energetic jet that helps to attach the flow against the adverse pressure gradient flow at the boundary layer of the diffuser [48,49]; it recovers the pressure gradient and prevents the stall [50].

The inlet boundary condition was considered velocity inlet with a constant value of 10 m/s and the outlet boundary condition was pressure outlet. The upper and lower boundary conditions were considered symmetry. To define rotational motion, the Sliding Mesh Motion (SMM) was considered in the inner cell zone. Boundary conditions and the computational domain are shown in Figure 3.

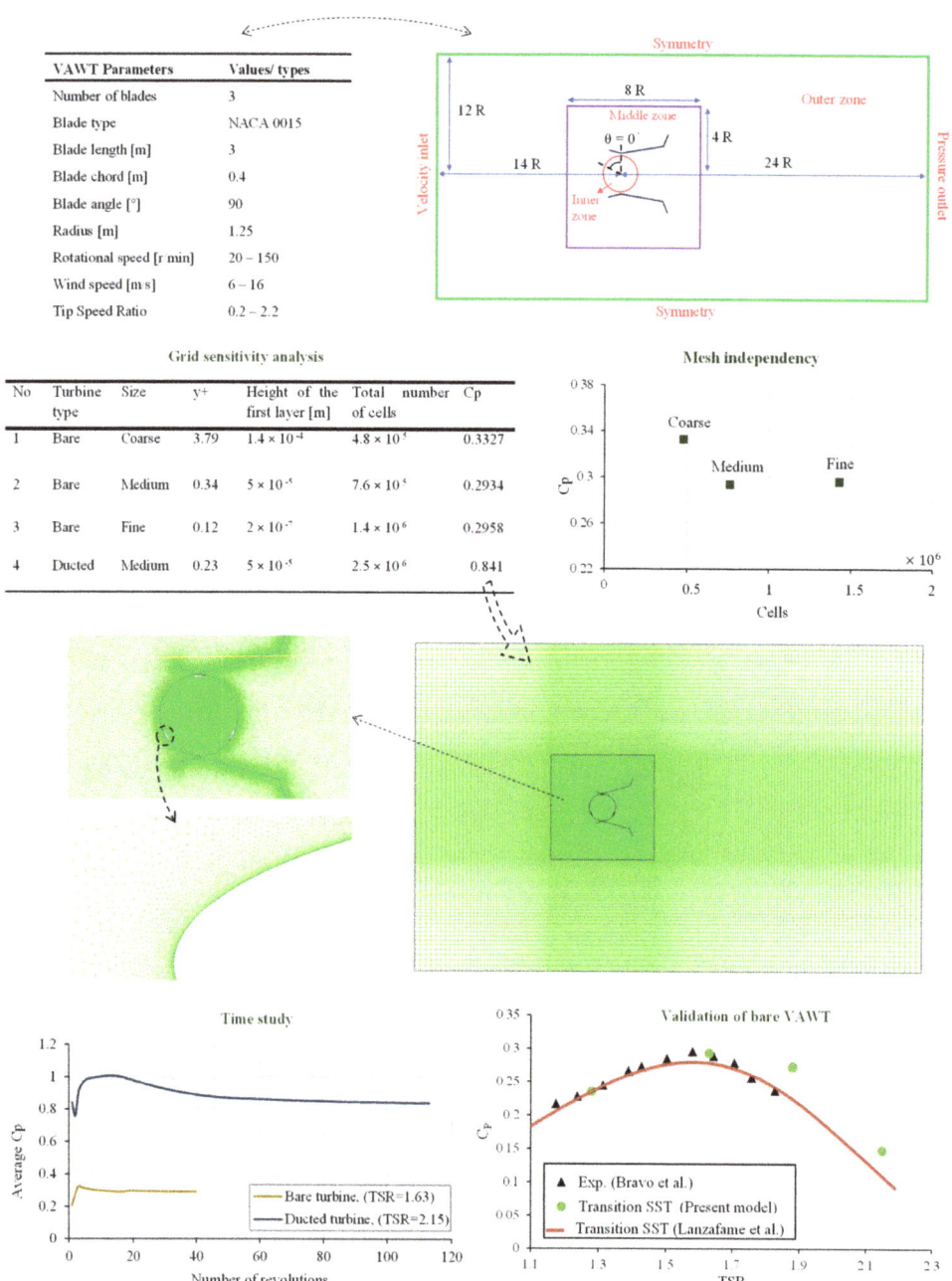

Figure 3. Numerical domain of ducted wind turbine and Bare wind turbine validation with the experimental data of Bravo et al. [47] and Transition SST results of Lanzafame et al. [51].

The two-dimensional, transient, and incompressible equations, the continuity equation derived from the conservation of mass law, Equation (1), and the Navier-Stokes equations, Equation (2), were used. Standard k-ε, k-ω and transition Shear Stress Transport (SST) turbulence models were tested by Lanzafame et al. [51]. Their results showed that the transition SST model has a good agreement with the experimental data, so this model was used in the present simulation.

The governing equations were calculated via Pressure-Implicit with Splitting of Operators (PISO). The second-order upwind method discretized the equations.

The flow domain of the ducted wind turbine was divided into inner, middle, and outer zones (Figure 3). The flow domain for the bare turbine had just the inner and outer zones. At first, mesh-sensitivity analysis for the bare turbine was done with three sizes, named as coarse, medium, and fine (Figure 3). The discrepancy between C_P of the coarse mesh and the fine mesh was 12.5%; this difference between the medium mesh and the fine mesh was decreased to less than 0.8%. In order to reduce the computational cost, the medium size mesh (7.6×10^5 cells) with less than 1% error was selected. The y+ value around the blade for the medium mesh was much less than 1. The medium mesh was also applied to the ducted wind turbine and showed the y+ value less than 1. For the inner zone, where the rotor was inserted, triangular meshes were generated. A boundary layer mesh with a growth rate of 1.1 from the walls of each blade in the interior zone was applied. This boundary layer consisted of 12 layers, with the first layer thickness of 5×10^{-5} m. The middle zone with a square boundary covered the duct. Triangular meshes with a growth rate of 1.05 were generated inside the middle zone. The outer zone was fully covered by quadratic structured meshes.

In Figure 3, the average power coefficients are plotted versus the number of cycles. The optimum number of cycles for the ducted turbine was about 85–90 and for the bare turbine was 28–30 cycles. The duct flow study showed that the velocity was oscillating in the throat because of the vortex shedding in the wake. In this regard, the number of revolutions required to stabilize for the ducted turbine was higher than that for the bare turbine, which agrees with the results of Zanforlin and Letizia [27]. Moreover, higher angular velocities of the rotor increased the number of required revolutions.

For the numerical verification, the current numerical results were compared with the experimental results provided by Bravo et al. [47], and the numerical transition SST results were presented by Lanzafame et al. [51]. The power coefficients, $C_P = P/(0.5\rho U_\infty^3 A)$, of the bare wind turbine versus the tip speed ratio (TSR) are shown in Figure 3. TSR is calculated as $TSR = (R\omega)/U_\infty$. U_∞ is the free stream velocity (or inlet/reference velocity), ω is the angular velocity, R is the rotor radius, P is extracted power, ρ is fluid density, and A is the rotor area. The comparison indicated that the current numerical results followed the trend of the experimental data of [50] and the numerical data of [51]. It showed that the transition SST turbulent model can capture the trends of the experimental data perfectly, which was also reported by [51].

5. Ducted Wind Turbine vs. Bare Wind Turbine

The aim of the study was the power coefficient increment of the VAWT using a duct. The power coefficients from the wind turbine were compared with those from the wind turbine located inside the optimized duct (Figure 4). The power coefficient obtained from the ducted VAWT was much higher than the bare one. The maximum value of the power coefficient 0.84 occurred at the TSR of 2.15. Using a duct, the extracted power of the wind turbines was improved in the other studies, too [22,25]. Watanabe et al. [22] studied some ducts with different geometry and showed a power coefficient increment by 2.6 times more than an open wind turbine. Here, the maximum power coefficient was enhanced up to 2.9 times and postponed to a higher TSR, which agrees with the study of Watanabe et al. [22]. For some places, such as free spaces between tall buildings, the direction of the wind is almost constant. Where the direction of the wind does not vary significantly using ducted VAWTs is recommended to generate more power.

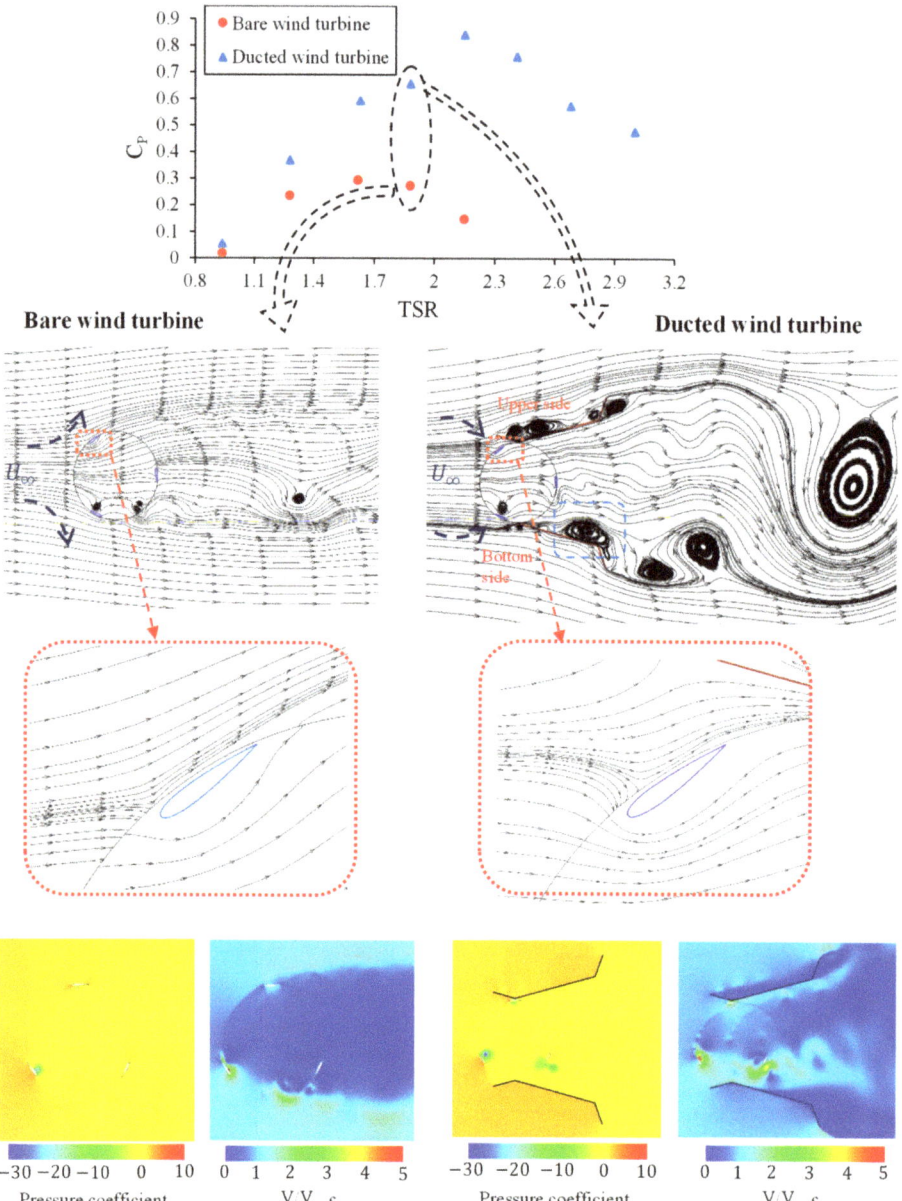

Figure 4. Performance of ducted wind turbine vs. bare wind turbine.

The inlet streamlines around the blades were platted at an azimuth angle of 30 degrees. The streamlines showed that the nozzle converged the input freestream flow and enhanced the incident velocity on each blade; the width of the stream-tube at the throat was decreased compared to that at the inlet. It was shown that the throat cross-section velocity was increased by 1.97 times when the duct was used. The dimensionless velocity and pressure coefficient contours for the ducted turbine and the bare turbine are shown in Figure 4. The flow velocity in the ducted turbine domain was higher than that in the bare turbine domain. More pressure coefficient differences between the sides of the blades also proved that the aerodynamic loads were increased for each blade when the rotor was covered by a duct. Therefore, the blades of the ducted turbine generated a higher torque than the blades of the bare turbine.

As shown in the streamlines of the ducted turbine in Figure 4, the walls of the duct caused vortex generation, such as the vortex marked with a blue box. When a rotor was inserted inside the duct, the horizontal and vertical components of the velocity were changed. As the rotor rotated, a low-pressure region was formed inside the bottom side of the diffuser causing an adverse pressure gradient; the flow separated, and some vortices were generated and shed to the wake periodically.

In Figure 5, the power-coefficient comparison of a single blade between the bare and ducted turbines showed that the ducted turbine experienced a higher power coefficient in the azimuth angles of 0–90 degrees. For VAWTs, each section of the blade experienced a different angle of attack (AOA) during one rotation of the turbine. For the bare wind turbine, the AOA was oscillating [52].

$$\alpha(\theta) = \arctan\left(\frac{\sin(\theta)}{TSR + \cos(\theta)}\right) \tag{3}$$

The AOA oscillation, Figure 5, resulted in dynamic stall (DS) phenomena. DS phenomena of wind turbines can be studied by an oscillating airfoil [28]. In this regard, besides the power extraction, the vortical structure of a blade can show valuable information for the rotor study. Since a leading-edge vortex is a very low-pressure vortex, the existence of the vortex on the suction side results in a high-pressure difference between the two sides of the airfoil and higher aerodynamic loads [26].

Figure 5 also shows the vorticity fields of the bare and ducted wind turbines at the TSR of 1.88. The flow structure from Figure 5a,b shows that the AOA of the airfoil was increasing through turbine rotation, which resulted in enhancing the lift and drag loads [28]. In Figure 5b, a small, leading-edge vortex is visible for the ducted turbine. The formation of the leading-edge vortex as the main characteristic of DS resulted in the power coefficient reduction of the blade. After 10° rotation of the rotor (Figure 5c), for the bare turbine, the first leading-edge vortex was seen, while the leading-edge vortex of the ducted one was more developed. There was also an approximately 10 degrees' azimuthal angle difference between the first maximum blade power coefficients of the bare and ducted blades, which agrees well with the leading-edge vortex generation. Thus, the duct caused advancing dynamic stall vortex generation of the blades. In Figure 5d,e, the leading-edge vortex development is visible for both turbines. Based on the developed leading-edge vortices, for both cases, the deep dynamic stall phenomena happened. In Figure 5f,g, the leading edge vortices are separated from the blades, while the AOA is decreasing. Figure 5h,i display that the airfoil experienced low angles with vortex sheets around the airfoil. Now, the dynamic stall loop was completed for both cases. Although the high wind speed through the throat passing the airfoil of the ducted turbine resulted in an improved power coefficient, the dynamic stall vortices including the leading-edge vortex caused power coefficient reduction. In sum, the wind turbine experienced a higher power coefficient when it was located inside the duct.

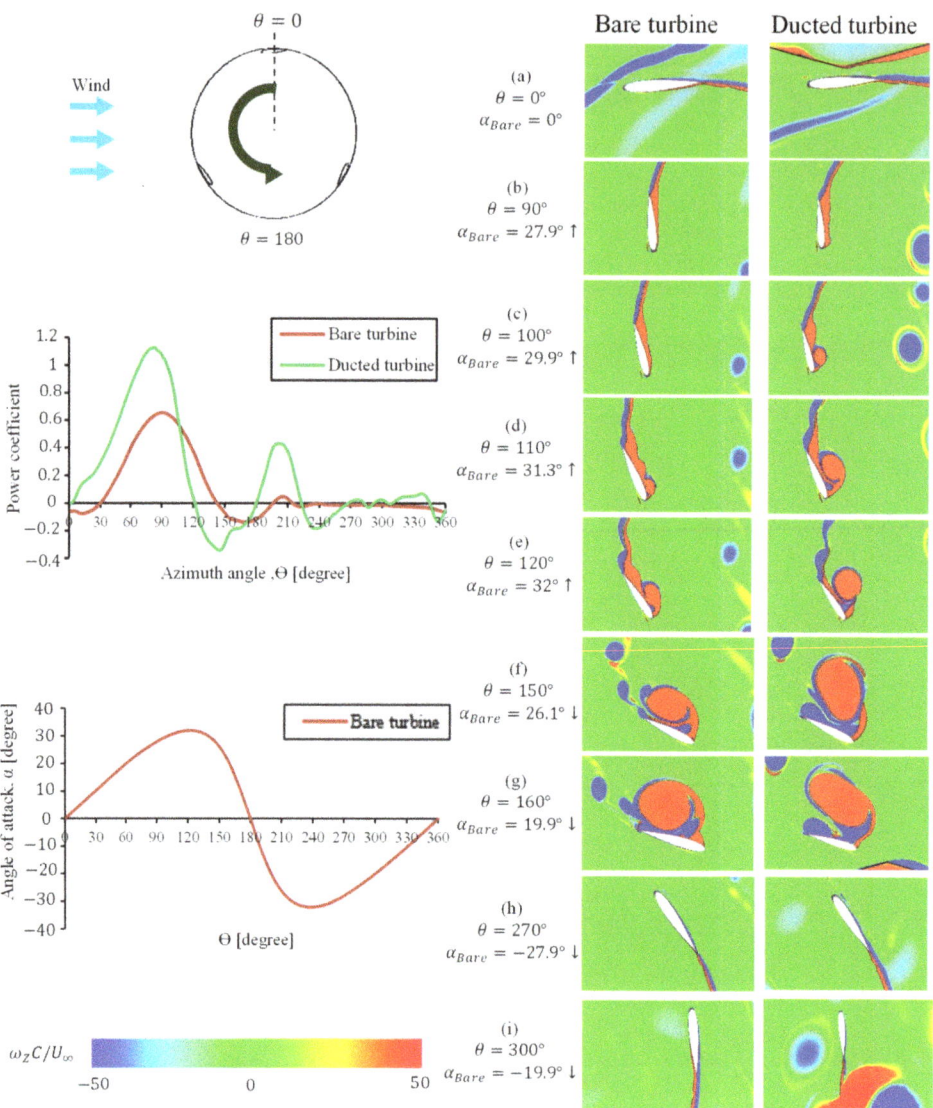

Figure 5. Power coefficient, angle of attack, and vorticity field of one blade in a cycle (*TSR* = 1.88).

6. Conclusions

For a DAWT, the optimum angles for three components of the duct consisting of the nozzle, the diffuser, and the flange were established. A code was developed whereby, for each case, a geometry was introduced, and a structured mesh was generated automatically; then each case was solved numerically. The DOE and Kriging methods were used to estimate the output parameters based on input variables. The maximum error of numerical results using the k-ω model in comparison with the experimental results was 5%. Finally, the optimum angles of 15, 15, and 70 degrees for the nozzle, the diffuser, and the flange, respectively, gave the highest flow velocity in the throat. The average velocity of the throat was increased by 1.97 times in the optimum duct. The velocity distribution in 90% of the throat area was almost uniform, having the maximum kinetic energy of the flow. Since the dimensionless values for the duct lengths are provided, the results may be used for other studies.

The second part of this study was related to placing the H-type rotor in the throat of the optimum duct. The maximum power coefficient reached 0.84 at the *TSR* of 2.15. The power coefficient of the ducted wind turbine was increased 2.9 times compared to that of the bare wind turbine. The *TSR* associated with the optimum power coefficient was shifted 0.52 units for the ducted wind turbine.

The vortical structure of the rotors revealed deep dynamic stall phenomena. The numerical simulation captured all the details of the phenomena. A comparison between the bare turbine and the ducted turbine showed that the dynamic stall leading-edge vortex generation was advanced and its separation was postponed when the rotor was located in the duct. Although dynamic stall resulted in declining performance of the turbine, the overall flow change inside the duct enhanced the generated power of the ducted wind turbine. It can be concluded that using a duct to cover the rotor of a wind turbine properly is an efficient way to increase the extracted power. The ducted VAWT is recommended for places with uniform wind directions.

Author Contributions: Methodology, validation, and discussion, M.H.R., B.R. and S.A.N.; supervision, K.G. discussion, review, and editing, K.G., M.S., A.A.-H., and J.N. All authors have read and agreed to the published version of the manuscript.

Funding: This research received no external funding.

Acknowledgments: The authors would like to thank Morteza Torabi for his assistance.

Conflicts of Interest: The authors declare no conflict of interest.

References

1. Anon, D. *Costa Head Experimental Wind Turbine*; Orkney Sustainable Energy Ltd.: Orkney, UK, 2010. Available online: http://www.orkneywind.co.uk/costa.html (accessed on 12 July 2021).
2. Lilley, G.; Rainbird, W. A Preliminary Report on the Design and Performance of Ducted Windmills. 1956. Available online: http://dspace.lib.cranfield.ac.uk/handle/1826/7971 (accessed on 12 July 2021).
3. Saleem, A.; Kim, M.-H. Aerodynamic analysis of an airborne wind turbine with three different airfoil-based buoyant shells using steady RANS simulations. *Energy Convers. Manag.* **2018**, *177*, 233–248. [CrossRef]
4. Bontempo, R.; Manna, M. On the potential of the ideal diffuser augmented wind turbine: An investigation by means of a momentum theory approach and of a free-wake ring-vortex actuator disk model. *Energy Convers. Manag.* **2020**, *213*, 112794. [CrossRef]
5. Saleem, A.; Kim, M.-H. Effect of rotor tip clearance on the aerodynamic performance of an aerofoil-based ducted wind turbine. *Energy Convers. Manag.* **2019**, *201*, 112186. [CrossRef]
6. Avallone, F.; Ragni, D.; Casalino, D. On the effect of the tip-clearance ratio on the aeroacoustics of a diffuser-augmented wind turbine. *Renew. Energy* **2020**, *152*, 1317–1327. [CrossRef]
7. Saleem, A.; Kim, M.-H. Performance of buoyant shell horizontal axis wind turbine under fluctuating yaw angles. *Energy* **2019**, *169*, 79–91. [CrossRef]
8. Suri, D.; Mukherjee, A.; Nayak, R.; Radhakrishnan, J.; Kwee, N.Y. Lighter-Than-Air Wind Energy Systems: Stability Analysis. In Proceedings of the International Conference on Mechanical Materials and Renewable Energy 2019, Rangpo, India, 6–7 December 2019.
9. Dilimulati, A.; Stathopoulos, T.; Paraschivoiu, M. Wind turbine designs for urban applications: A case study of shrouded diffuser casing for turbines. *J. Wind Eng. Ind. Aerodyn.* **2018**, *175*, 179–192. [CrossRef]
10. Bontempo, R.; Manna, M. Diffuser augmented wind turbines: Review and assessment of theoretical models. *Appl. Energy* **2020**, *280*, 115867. [CrossRef]
11. Van Bussel, G.J. The science of making more torque from wind: Diffuser experiments and theory revisited. *J. Phys. Conf. Ser.* **2007**, *75*, 012010. [CrossRef]
12. Bontempo, R.; Manna, M. Effects of the duct thrust on the performance of ducted wind turbines. *Energy* **2016**, *99*, 274–287. [CrossRef]
13. Igra, O. Research and development for shrouded wind turbines. *Energy Convers. Manag.* **1981**, *21*, 13–48. [CrossRef]
14. Agha, A.; Chaudhry, H.N.; Wang, F. Diffuser augmented wind turbine (DAWT) technologies: A review. *Int. J. Renew. Energy Res. (IJRER)* **2018**, *8*, 1369–1385.

15. Ohya, Y.; Karasudani, T. A shrouded wind turbine generating high output power with wind-lens technology. *Energies* **2010**, *3*, 634–649. [CrossRef]
16. Heikal, H.A.; Abu-Elyazeed, O.S.; Nawar, M.A.; Attai, Y.A.; Mohamed, M.M. On the actual power coefficient by theoretical developing of the diffuser flange of wind-lens turbine. *Renew. Energy* **2018**, *125*, 295–305. [CrossRef]
17. Rochman, M.N.; Nasution, A.; Nugroho, G. CFD Studies on the Flanged Diffuser Augmented Wind Turbine with Optimized Curvature Wall. In *ICoSI 2014*; Springer: Singapore, 2017; pp. 347–355.
18. Ranjbar, M.H.; Nasrazadani, S.A.; Gharali, K. Optimization of a Flanged DAWT Using a CFD Actuator Disc Method. In *International Conference on Applied Mathematics, Modeling and Computational Science*; Springer: Waterloo, ON, Canada, 2017; pp. 219–228.
19. El-Zahaby, A.M.; Kabeel, A.; Elsayed, S.; Obiaa, M. CFD analysis of flow fields for shrouded wind turbine's diffuser model with different flange angles. *Alex. Eng. J.* **2017**, *56*, 171–179. [CrossRef]
20. Kosasih, B.; Tondelli, A. Experimental study of shrouded micro-wind turbine. *Procedia Eng.* **2012**, *49*, 92–98. [CrossRef]
21. Wang, X.; Wong, K.; Chong, W.; Ng, J.; Xiang, X.; Wang, C. Experimental investigation of a diffuser-integrated vertical axis wind turbine. In *IOP Conference Series: Earth and Environmental Science*; IOP: Bangkok, Thailand, 2020.
22. Watanabe, K.; Takahashi, S.; Ohya, Y. Application of a diffuser structure to vertical-axis wind turbines. *Energies* **2016**, *9*, 406. [CrossRef]
23. Ghazalla, R.; Mohamed, M.; Hafiz, A. Synergistic analysis of a Darrieus wind turbine using computational fluid dynamics. *Energy* **2019**, *189*, 116214. [CrossRef]
24. Zanforlin, S.; Buzzi, F.; Francesconi, M. Performance Analysis of Hydrofoil Shaped and Bi-Directional Diffusers for Cross Flow Tidal Turbines in Single and Double-Rotor Configurations. *Energies* **2019**, *12*, 272. [CrossRef]
25. Hashem, I.; Mohamed, M. Aerodynamic performance enhancements of H-rotor Darrieus wind turbine. *Energy* **2018**, *142*, 531–545. [CrossRef]
26. Dessoky, A.; Bangga, G.; Lutz, T.; Krämer, E. Aerodynamic and aeroacoustic performance assessment of H-rotor darrieus VAWT equipped with wind-lens technology. *Energy* **2019**, *175*, 76–97. [CrossRef]
27. Zanforlin, S.; Letizia, S. Effects of upstream buildings on the performance of a synergistic roof-and-diffuser augmentation system for cross flow wind turbines. *J. Wind Eng. Ind. Aerodyn.* **2019**, *184*, 329–341. [CrossRef]
28. Gharali, K.; Johnson, D.A. Dynamic stall simulation of a pitching airfoil under unsteady freestream velocity. *J. Fluids Struct.* **2013**, *42*, 228–244. [CrossRef]
29. Gharali, K.; Johnson, D.A.; Lam, V.; Gu, M. A 2D blade element study of a wind turbine rotor under yaw loads. *Wind Eng.* **2015**, *39*, 557–567. [CrossRef]
30. Gharali, K.; Gharaei, E.; Soltani, M.; Raahemifar, K. Reduced frequency effects on combined oscillations, angle of attack and free stream oscillations, for a wind turbine blade element. *Renew. Energy* **2018**, *115*, 252–259. [CrossRef]
31. Bakhtiari, E.; Gharali, K.; Chini, S.F. Corrigendum to "Super-hydrophobicity effects on performance of a dynamic wind turbine blade element under yaw loads". *Renew. Energy* **2019**, *140*, 539–551, Erratum in: **2020**, *147*, 2528. [CrossRef]
32. Hamada, K.; Smith, T.; Durrani, N.; Qin, N.; Howell, R. Unsteady flow simulation and dynamic stall around vertical axis wind turbine blades. In Proceedings of the 46th AIAA Aerospace Sciences Meeting and Exhibit, Reno, NV, USA, 7–10 January 2008; p. 1319.
33. Bangga, G.; Hutomo, G.; Wiranegara, R.; Sasongko, H. Numerical study on a single bladed vertical axis wind turbine under dynamic stall. *J. Mech. Sci. Technol.* **2017**, *31*, 261–267. [CrossRef]
34. Jamieson, P.M. Beating betz: Energy extraction limits in a constrained flow field. *J. Sol. Energy Eng.* **2009**, *131*. [CrossRef]
35. Patankar, S.V. *Numerical Heat Transfer and Fluid Flow (Book)*; Hemisphere Publishing Corp.: Washington, DC, USA, 1980; 210p.
36. Ohya, Y.; Karasudani, T.; Sakurai, A.; Abe, K.-I.; Inoue, M. Development of a shrouded wind turbine with a flanged diffuser. *J. Wind Eng. Ind. Aerodyn.* **2008**, *96*, 524–539. [CrossRef]
37. Bontempo, R.; Manna, M. Performance analysis of open and ducted wind turbines. *Appl. Energy* **2014**, *136*, 405–416. [CrossRef]
38. Khamlaj, T.A.; Rumpfkeil, M.P. Analysis and optimization of ducted wind turbines. *Energy* **2018**, *162*, 1234–1252. [CrossRef]
39. Abe, K.-i.; Ohya, Y. An investigation of flow fields around flanged diffusers using CFD. *J. Wind Eng. Ind. Aerodyn.* **2004**, *92*, 315–330. [CrossRef]
40. Box, G.; Wilson, K. On the Experimental Attainment of Optimum Conditions. *J. R. Stat. Society. Ser. B (Methodol.)* **1951**, *13*, 1–45. [CrossRef]
41. Cavazzuti, M. *Optimization Methods: From Theory to Design Scientific and Technological Aspects in Mechanics*; Springer Science & Business Media: Berlin, Germany, 2012.
42. Dimitrov, N.; Kelly, M.C.; Vignaroli, A.; Berg, J. From wind to loads: Wind turbine site-specific load estimation with surrogate models trained on high-fidelity load databases. *Wind Energy Sci.* **2018**, *3*, 767–790. [CrossRef]
43. Kumar, P.M.; Seo, J.; Seok, W.; Rhee, S.H.; Samad, A. Multi-fidelity optimization of blade thickness parameters for a horizontal axis tidal stream turbine. *Renew. Energy* **2019**, *135*, 277–287. [CrossRef]
44. Slot, R.M.; Sørensen, J.D.; Sudret, B.; Svenningsen, L.; Thøgersen, M.L. Surrogate model uncertainty in wind turbine reliability assessment. *Renew. Energy* **2020**, *151*, 1150–1162. [CrossRef]
45. Ranjbar, M.H.; Nasrazadani, S.A.; Zanganeh Kia, H.; Gharali, K. Reaching the betz limit experimentally and numerically. *Energy Equip. Syst.* **2019**, *7*, 271–278.

46. Ranjbar, M.H.; Zanganeh Kia, H.; Nasrazadani, S.A.; Gharali, K.; Nathwani, J. Experimental and numerical investigations of actuator disks for wind turbines. *Energy Sci. Eng.* **2020**, *8*, 2371–2386. [CrossRef]
47. Bravo, R.; Tullis, S.; Ziada, S. Performance testing of a small vertical-axis wind turbine. In Proceedings of the 21st Canadian Congress of Applied Mechanics, Toronto, ON, Canada, 3–7 June 2007.
48. Hansen, M.O.L.; Sørensen, N.N.; Flay, R. Effect of placing a diffuser around a wind turbine. *Wind Energy Int. J. Prog. Appl. Wind Power Convers. Technol.* **2000**, *3*, 207–213. [CrossRef]
49. Cresswell, N.; Ingram, G.; Dominy, R. The impact of diffuser augmentation on a tidal stream turbine. *Ocean Eng.* **2015**, *108*, 155–163. [CrossRef]
50. Kwong, A.H.; Dowling, A.P. Active boundary-layer control in diffusers. *AIAA J.* **1994**, *32*, 2409–2414. [CrossRef]
51. Lanzafame, R.; Mauro, S.; Messina, M. 2D CFD modeling of H-Darrieus wind turbines using a transition turbulence model. *Energy Procedia* **2014**, *45*, 131–140. [CrossRef]
52. Paraschivoiu, I. *Wind Turbine Design: With Emphasis on Darrieus Concept*; Polytechnique: Montreal, QC, Canada, 2002.

Article

Piloted Simulation of the Rotorcraft Wind Turbine Wake Interaction during Hover and Transit Flights

Alexander Štrbac [1,*], Daniel Heinrich Greiwe [1], Frauke Hoffmann [1], Marion Cormier [2] and Thorsten Lutz [2]

1. German Aerospace Center (DLR), Institute of Flight Systems, 38108 Braunschweig, Germany; daniel.greiwe@dlr.de (D.H.G.); frauke.hoffmann@dlr.de (F.H.)
2. Institute of Aerodynamics and Gas Dynamics (IAG), University of Stuttgart, 70569 Stuttgart, Germany; marion.cormier@iag.uni-stuttgart.de (M.C.); thorsten.lutz@iag.uni-stuttgart.de (T.L.)
* Correspondence: Alexander.Strbac@dlr.de

Abstract: Helicopters are used for offshore wind farms for maintenance and support flights. The number of helicopter operations is increasing with the expansion of offshore wind energy, which stresses the point that the current German regulations have not yet been validated through scientific analysis. A collaborative research project between DLR, the Technical University of Munich, the University of Stuttgart and the University of Tübingen has been conducted to examine the sizes of the flight corridors on offshore wind farms and the lateral safety clearance for helicopter hoist operations at offshore wind turbines. This paper details the results of piloted helicopter simulations in a realistic offshore wind farm scenario. The far-wake of rotating wind turbines and the near-wake of non-rotating wind turbines have been simulated with high-fidelity computational fluid dynamics under realistic turbulent inflow conditions. The resulting flow fields have been processed by superposition during piloted simulations in the research flight simulator AVES to examine the flight corridors in transit flights and the lateral safety clearance in hovering flights. The results suggest a sufficient size for the flight corridor and sufficient lateral safety clearance at the offshore wind turbines in the considered scenarios.

Keywords: handling qualities; piloted simulation; wind turbine wake; helicopter vortex–rotor interaction; wake vortex encounter; helicopter offshore operation; flight safety; rotorcraft; computational fluid dynamics; offshore wind energy

1. Introduction

Europe's efforts to develop sustainable and affordable energy production are leading to the rapid expansion of offshore wind energy. Among other means, helicopters are used at offshore wind farms for maintenance and support flights, and the number of helicopter operations is increasing along with the expansion of offshore wind energy. Current German regulations for helicopter operations in offshore wind farms to protect helicopters from potentially dangerous wakes of wind turbines (WTs) underlie assumptions, which are not scientifically sound. Therefore, the HeliOW project (helicopter offshore wind) has been established to assess the suitability of the German regulations to protecting helicopters form potential safety risks. It is a collaborative national research project which includes in situ measurements of WT wakes with unmanned aircraft systems (University of Tübingen [1]), high-fidelity computational fluid dynamic (CFD) simulations of WT wakes (University of Stuttgart [2–5]), desktop helicopter simulations with mutual interaction between WT wake and helicopter (Technical University of Munich [6,7]) and piloted helicopter simulations in a research flight simulator with a superposition method (DLR [8]).

Since 2015, the average nominal power of newly installed offshore WTs has grown at an annual rate of 16% [9]. Currently, wind farms are located up to 100 km from shore in water up to 100 m deep by using bottom-fixed and floating technologies [9]. Current wind farms consist of up to 165 offshore WTs, and operational offshore WTs with nominal power generation of up to 9.5 MW can be found [9]. Newly ordered offshore WTs have reached an average nominal power lvel of 10.4 MW [10], and manufacturers have announced the development of future offshore WTs above 14 MW (https://www.ge.com/renewableenergy/wind-energy/offshore-wind/haliade-x-offshore-turbine (accessed on 21 September 2021)) or 15 MW (https://www.vestas.com/en/products/offshore%20platforms/v236_15_mw#!) (accessed on 23 September 2021).

A typical offshore wind farm consists of a number of WTs and a manned offshore substation (OSS). The OSS is used for maintenance of the wind farm and is usually located at its center. Wind farm operators use crew transfer vessels (CTVs) or helicopters to transport maintenance engineers, tools spare parts from the OSS to inoperative WTs. CTVs offer high passenger and cargo capacity, but they are typically limited to a maximum of sea state 4 ([11], Table A3) and passengers may be affected by seasickness. The benefits of helicopters are short transfer times and an operational limit of sea state 6, based on their rotorcraft flotation systems ([12], Table A3). Therefore, a helicopter is most beneficial for urgent issues (unscheduled WT maintenance, high sea states, emergency transport, etc.) and for wind farms located far from shore.

Potential risks for helicopter operations originate from bluff-body wakes in proximity of structures (OSS, WTs) and from the wakes of rotating WTs (turbulence, wind deficit, blade tip vortices). The latter may cause a so-called helicopter vortex–rotor interaction with blade tip vortices from the WTs. The influence of vortex–rotor interactions of helicopters with fixed-wing aircraft vortex wakes has been examined in the past by piloted simulations [13]. Those results are not directly transferable to WT wakes, but a promising subjective pilot rating scale has been developed.

The longitudinal vortex–rotor interaction between helicopters and WTs has been analyzed by van der Wall et al. [14]. A semi-empirical wind turbine wake model (SWM) has been developed to assess the influence of the WT vortex on a Bo105 rotor trim. The location of the vortex within the rotor disk, the nominal WT power and the distance between helicopter and WT have been varied. These examinations have been extended to examine the influences of various rotor sizes, blade flapping motions and advance ratios [15]; different vortex orientations within the rotor disk [16]; and vortex deflection [17]. Lastly, the modeling approaches for vortex–rotor interactions with a rigid line vortex and a deflected line vortex have been compared with intermediate-fidelity free-wake simulations, and guidelines for their applicability are given in [18]. Overall, these investigations give insights into the complicated effects at the helicopter's main rotor. However, these results are limited to rotor trim, and flight dynamics are neglected.

Furthermore, an overview of Europe research on the influences of onshore/offshore WT wakes on helicopter operations is given in a report compiled by members of the GARTEUR Helicopter Action Group 23 (HC-AG23 [19]). It addresses different topics, such as WT wake experiments and computations, offline helicopter simulation, and piloted helicopter simulations. However, the results of the piloted simulations are very limited and show the need for further examinations.

The first extensive piloted simulation of transit flights in offshore farms has been performed by the authors of [8], based on flow fields from the SWM and subjective pilot rating scales. WT vortex encounters of a EC135 helicopter without any stability systems (e.g., SAS) have been examined for different airspeeds and WTs with a nominal power up to 20 MW. Overall, the results suggest sufficient safety clearance. However, the suitability of the SWM has not been proven.

This study deepens the previous efforts made by the authors by using flow fields from high-fidelity CFD instead of flow fields from the SWM for transit flights. Furthermore, a second operational scenario has been derived from helicopter hoist operations at offshore WTs and has been examined by piloted simulations. This paper proceeds as follows. Firstly, the current German regulations for helicopter operations in offshore wind farms are briefly described in regard to the flight corridor within a wind farm and the lateral safety distance for helicopter hoist operations at offshore WTs. Secondly, two typical operational scenarios for helicopters are derived from operational practice for the assessment of potential risks. Thirdly, the generation of WT wake flow fields in both operational scenarios from CFD is described and the resulting flow fields are analyzed. Fourthly, the setup, performance and assessment of the piloted simulations are described. Fifthly, the results of both operational scenarios are analyzed. In addition, results of offline simulations to compare WT wake flow fields from the previously used SWM and high-fidelity CFD are given. Finally, all results are discussed and a conclusion is given.

2. Regulations for Helicopter Operations

Maritime helicopter operations in offshore wind farms involve the transportation of maintenance engineers from the mainland to the OSS or from the OSS to single WTs. The former represents a transit flight from outside through the wind farm and a landing on the OSS. The latter represents a short distance flight to a single WT followed by a helicopter hoist operation of persons, tools and spare parts. Both helicopter operations are performed in an adverse environment with changing weather conditions, low visibility, obstacles and turbulence.

German authorities have defined regulations for an inner and outer flight corridor used for transit flights to the OSS (Figure 1a). Its dimensions are based on the geometry of WTs and empirical experience [20]. In particular, the rotor radius of the WT is used as a scaling factor, which causes an increase in the dimensions of the inner and outer flight corridor for future WTs. The influences of wind speed, its direction and the WT wake are neglected.

Furthermore, German authorities have defined regulations for helicopter hoist operations at offshore WTs [21]. An arbitrary hoist area at the top of the nacelle is required, which contains a square of 4×4 m as the minimum size (Figure 1b). It includes a hoist position, which is centered over the hoist area. The minimum lateral safety clearance is defined for the hoist maneuver directly above the hoist position. The distance between the helicopter rotor disk and the rear of the WT rotor disk must exceed 5 m. However, it is recommended to choose a reference helicopter and increase lateral safety clearance to half of the rotor diameter D_H. Additional requirements apply for obstacles and railings and are not further discussed here. Overall, the regulations simply recommend the rotor diameter D_H as a scaling factor. The influences of wind speed and the wakes of surrounding WTs are neglected as well.

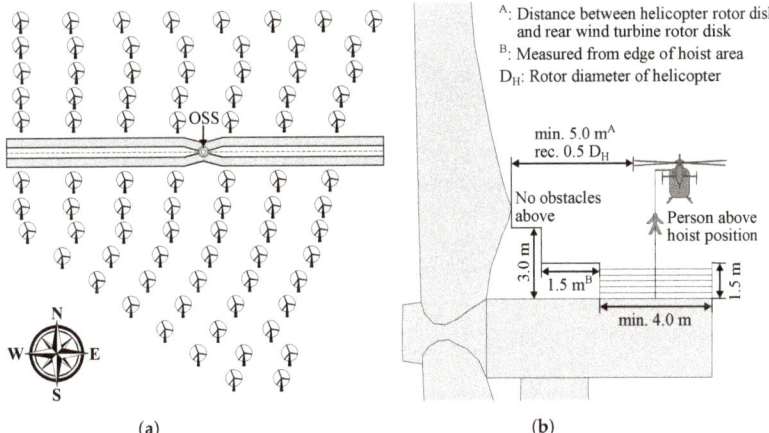

Figure 1. German regulations for maritime helicopter operations in offshore wind farms. (a) The inner and outer flight corridor at the offshore wind farm Global Tech I (GTI) [20]; (b) Lateral safety clearance for the helicopter hoist area at offshore wind turbines (WTs) [21].

3. Operational Scenarios

After consultations with the Federal Aviation Office of Germany (LBA) and the Federal Maritime and Hydrographic Agency of Germany (BSH), two potentially critical operational scenarios for the piloted simulation were derived.

Operational Scenario 1 (OS-1) describes a transit flight in the flight corridor in proximity of rotating WTs (Figure 2) and was used to examine the size of the flight corridor (Figure 1a). A potential risk is the wake of rotating WTs, which is convected downstream by the wind and crosses the flight corridor. It contains a wind deficit surrounded by helical WT blade tip vortices, which are characterized by high velocity gradients. As depicted in Figure 2, the local vortex axis and the longitudinal helicopter axis can coincide, which causes so-called longitudinal vortex rotor interactions with primary excitation of the helicopter pitch axis [16]. For transit flights at WT hub height, so-called orthogonal vortex rotor interactions can occur [16].

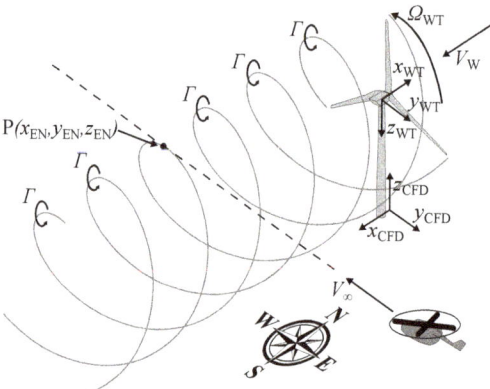

Figure 2. Operational Scenario 1 (OS-1): Transit flight with direction west next to a rotating WT.

Operational Scenario 2 (OS-2) describes a hover flight at a non-rotating WT (Figure 3) and was used to examine the lateral safety clearance at the helicopter hoist area (Figure 1b). It was derived from a typical helicopter hoist maneuver at an inoperative WT which needs to be maintained by engineers. The WT nacelle is perpendicular to the incoming wind V_W. The WT rotor is stopped in the so-called "L-position", resulting in one WT rotor blade facing horizontally into the incoming wind V_W. In operational practice, this position is preferred by helicopter pilots, because it allows the helicopter to fly into a headwind and to use the WT rotor blade as a visual reference. During approach and hovering at the WT, the pilot is guided by a hoist operator via radio to the helicopter hoist position. The hoist operator is placed at the helicopter winch and estimates visually the distance to the hoist position. Potential risks during hover are the bluff-body wake of the WT nacelle, proximity to obstacles and wakes from neighboring WTs. Note that the WT and CFD coordinate system have changed compared to OS-1 (Figure 3).

The OS-2 is split into two subcases. The first subcase represents a hover flight at an isolated WT, which is not affected by surrounding WTs (Figure 3a). Therefore, the incoming wind V_W represents typical maritime wind conditions without additional turbulence. This is typical for inoperative WTs at the edge of offshore wind farms. The second subcase represents a hover flight at a non-isolated WT, which is affected by surrounding WTs (Figure 3b). Therefore, the incoming wind V_W contains the turbulent wake of one single rotating WT, which is typical for inoperative WTs within offshore wind farms.

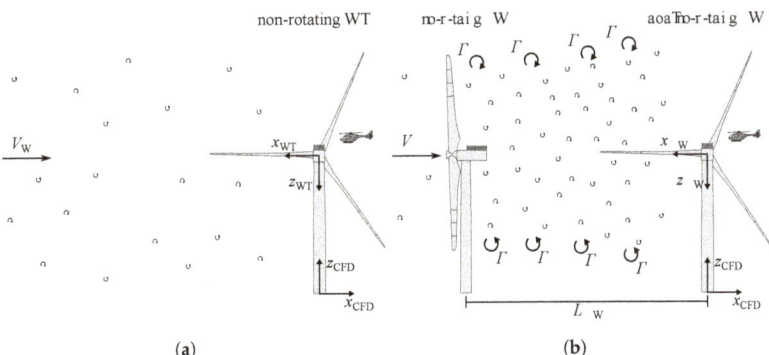

Figure 3. Operational Scenario 2 (OS-2): Hover flight at a non-rotating WT in "L-position". (**a**) Subcase 1: Hover flight at an isolated WT. (**b**) Subcase 2: Hover flight at a non-isolated WT ($L_{WT} = 680$ m).

Two potentially critical wind speeds V_W based on the OFFWINDTECH WT were chosen for OS-1 and OS-2. The OFFWINDTECH WT is a slightly modified version of the NREL 5 MW generic research WT [22] and resembles the real existing WTs in the offshore wind farm GTI. The first wind speed V_W corresponds to the expected strong blade tip vortices. WTs reach the maximum initial blade tip vortex circulation Γ_0 in the region of the rated wind speed $V_{W,\Omega max}$ [8], which is reached at a medium wind speed of $V_W = 11.3$ m/s for the considered WT. The other potentially critical wind speed was chosen as $V_W = 25.0$ m/s to represent extreme weather conditions. As helicopter hoist operations are performed until wind speeds V_W between 18–21 m/s due to limitations of the wind farm operators, the cut-off wind speed of the considered WT of $V_{W,max} = 25.0$ m/s was chosen as a conservative approach. Figure 4 shows the relative frequency of the wind speed V_W at the German Bight. As indicated, the chosen wind speeds V_W represent a common medium wind speed and a very high wind speed.

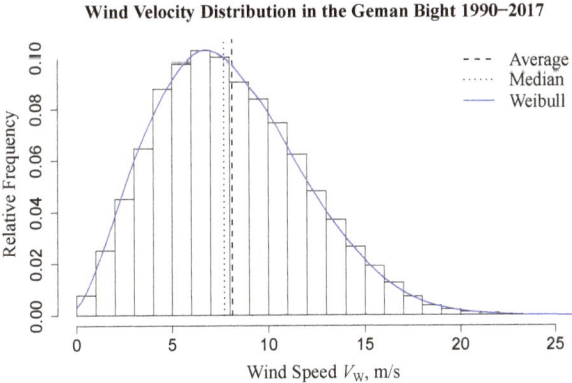

Figure 4. Measured wind speed V_W at a unmanned lightvessel named "Deutsche Bucht", located approximately 35 km north from the island Langeoog (Database from [23]).

As a result, two cases for OS-1 and four cases for OS-2 have been derived. Each case needs specific WT wake flow fields obtained from CFD for the helicopter simulation, which depends on WT status (rotating, non-rotating), WT surrounding (isolated, non-isolated) and wind speed V_W (11.3 m/s, 25.0 m/s). A summary of the postprocessed airwake datasets is depicted in Table 1, each dataset being denoted with a name from C1 to C7. Note that based on C1, an additional airwake dataset, C3, was defined to examine the influences of the maritime inflow data (Table 2) on the length and stability of the WT blade tip vortex helix, but was discarded beforehand and does not appear in Table 1. Furthermore, airwake dataset C6 was defined and scheduled for the piloted simulations, but was been examined due to time constraints.

4. Computational Fluid Dynamics

The WT wake flow fields were generated by performing time-resolved CFD computations on a high-performance computing cluster. Those were performed on structured meshes with local grid refinement and large spatial domains. In order to be processed in real-time by the helicopter flight simulator, this large amount of data had to be reduced to fit the simulator's memory constraints. Therefore, small spatial areas of interest were extracted from the original CFD data and interpolated on an equidistant structured mesh without local grid refinement. Afterwards, the resulting postprocessed airwake data were used as a lookup table in the piloted simulation.

Table 1 shows an overview of the interpolated airwake data. OS-1 considers transit helicopter flights with an airspeed of $V_\infty = 80\,\text{kt} \approx 41.2\,\text{m/s}$, which is much faster than the convection wind speed of the WT wake of $V_W = 10.3\,\text{m/s}$ and $V_W = 25.0\,\text{m/s}$. Therefore, the temporal evolution of the WT wake was neglected, and only steady airwake data (time-independent) were used for this case. Consequently, benefits of the reduced memory consumption have been used for large spatial dimensions (X_{WT}, Y_{WT}, Z_{WT}) with a fine spatial discretization (Δx_{WT}, Δy_{WT}, Δz_{WT}).

In contrast, the helicopter hover position remained the same for OS-2. Therefore, unsteady airwake data (time-dependent) had to be used with a temporal discretization Δt_{WT} and a duration T_{WT}. For the unsteady airwake data, every 8th (C4) or every 12th (C5, C7) time step was extracted from the original WT wake flow field. A smooth time loop within the helicopter simulation was used for the airwake data to allow continuous simulations. As a consequence, spatial dimensions (X_{WT}, Y_{WT}, Z_{WT}) were decreased and spatial discretization (Δx_{WT}, Δy_{WT}, Δz_{WT}) was coarsened compared to OS-1 due to memory consumption.

Furthermore, OS-1 and OS-2 have very different demands on the CFD methodology. OS-1 takes place at the far-wake of a rotating WT and is dependent on accurate fluid mechanical simulation of the WT blade tip vortices. In contrast, OS-2 takes place at the near-wake of a non-rotating WT and is dependent on accurate fluid mechanical simulation of the bluff-body aerodynamics associated with the flow around the nacelle.

Table 1. Overview of interpolated airwake data.

Airwake Data	C1	C2	C4	C5	C6	C7
Operational scenario	OS-1	OS-1	OS-2	OS-2	OS-2	OS-2
WT status	rotating	rotating	non-rotating	non-rotating	non-rotating	non-rotating
WT surrounding	isolated	isolated	isolated	isolated	non-isolated	non-isolated
V_W, m/s	11.3	25.0	11.3	25.0	11.3	25.0
Δx_{WT}, Δy_{WT}, Δz_{WT}, m	0.125	0.125	0.225	0.225	0.225	0.225
X_{WT}, m	440	440	85	85	85	85
Y_{WT}, m	232	232	170	170	170	170
Z_{WT}, m	206	206	50	50	50	50
Δt_{WT}, s	-	-	0.04	0.04	0.04	0.04
T_{WT}, s	-	-	12.40	12.40	12.40	12.40

All CFD flow fields around the WT were simulated with the finite volume flow solver FLOWer [24], which was originally developed by DLR and is continuously furthered by the University of Stuttgart for wind-energy applications. It is a block-structured solver, and the Chimera overlapping technique is used to connect the structured grids of the WT. Dual time-stepping and multigrid algorithms are applied to accelerate the numerical convergence. The 2nd order WENO scheme, the Jameson–Turkel–Schmidt (JST) scheme [25] or the 5th-order the WENO scheme [26] (spatial schemes) can be assigned block-wise in the computational domain to optimize computational time and resolution of vortical structures. The WENO scheme can better resolve the vortical structures and was thus applied in the regions of interest, except in the boundary layers regions where it was not applicable due to the high aspect ratio of the computational cells, and thus the JST scheme was applied instead.

The WT chosen for this study is the OFFWINDTECH WT, which was derived from the NREL 5 MW generic research WT and optimized for offshore conditions [27]. The main changes concern the rated operating point, with an inflow wind speed and rotational speed as defined below, and a blade pitch angle of $-2.29°$. It strongly resembles the real existing WTs in the offshore wind farm GTI. It features a rotor diameter of $D_{WT} = 126$ m and a rotational frequency of $f_{WT} = 11.7$ 1/min at the rated wind speed of $V_{W,\Omega \max} = 11.3$ m/s.

4.1. Rotating WT

In this section, the computational setup and resulting flow fields of OS-1 are presented, which were postprocessed afterwards to airwake data C1 and C2 (Table 1).

4.1.1. CFD Setup

As OS-1 depends on the accurate reproduction of the dynamics of the blade tip vortices downstream of the WT, the computational demands regarding spatial resolution and transport of the vortex are relatively high. The grid requirements for the reproduction of measured viscous vortex cores in CFD simulations with FLOWer were examined in [3]. In this study, good agreement between CFD results and UAV-based in situ measurements of the blade tip vortex characteristics downstream of a full-size commercial WT was found on the fine-resolved blade tip vortex grid. Based on these findings, the grid requirement in the wake of the considered WT was estimated to $\Delta x_{CFD} = 0.125$ m in the blade tip vortex region, taking into account geometrical scaling between the two WTs. The refined area around the blade tip vortices, which positions were determined based on a precursor simulation, is visualized in Figure 5.

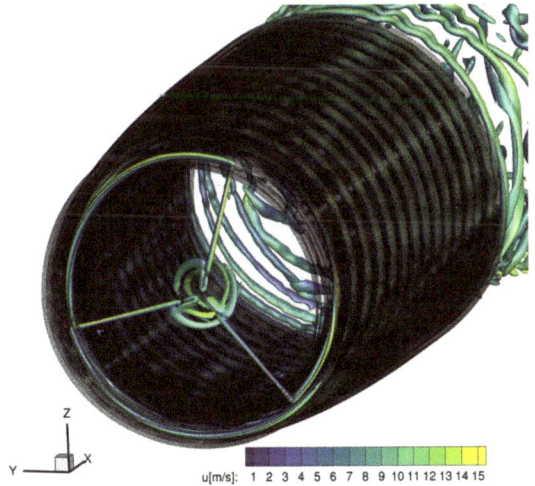

Figure 5. Grid refinement area around the blade tip vortex helix (OS-1).

For the present study, the detailed aerodynamics of the rotor itself are of little relevance, and only the effects of tower and nacelle that have a footprint in the near-wake of the WT and in the lower part of the wake are detailed. Thus, the actuator line approach (ACL) was used to model the WT, which has proven to appropriately model the wake dynamics [28,29]. Moreover, it allows one to save computational cells in the rotor area, which can then be reinvested in the wake region. Ninety-nine actuator line elements were introduced along the blade, with a higher density at the blade tip and root region. The same polar data were used as in [2] and were generated with Xfoil. The size of the ACL regularization kernel ϵ was set proportionally to the local airfoil chord length c_{WT}, $\epsilon = 0.25\, c_{WT}$, according to [30]. Moreover, the well-established numerical stability criterion $\epsilon \geq 2\, \Delta x_{CFD}$ [31] was fulfilled at every point.

Synthetic inflow data were used to recreate the turbulent inflow representative of offshore conditions in the computational setup at the wind speed at hub height of $V_W = 11.3\,\text{m/s}$ and $V_W = 25.0\,\text{m/s}$. The mean vertical wind profile was generated via the power law with the exponent $\alpha = 0.14$, which is representative of maritime conditions [32], and was imposed at the inlet via a Dirichlet boundary condition. Synthetic turbulent fluctuations were superposed onto this mean profile and modeled via the so-called Mann–Box approach [33], whereby they are added as volume forces downstream of the inlet plane. The turbulence statistics used as input parameters in the model were extracted from the literature data at the hub height $z_{HUB} = 90\,\text{m}$ and are gathered in Table 2. The turbulence intensity (I) and integral length scale (L) were determined based on the large database contained in [34], except for $I = 6.84\%$ at $V_W = 25.0\,\text{m/s}$, which was extracted from the FINO1 Offshore measurement campaign [32]. No thermal stratification effects are considered in the present study.

Far-field boundary conditions were applied at the outlet, the top and the lateral boundaries of the computational domain, and a non-slip wall was placed at the bottom. At the inlet, the mean wind velocity profile was imposed, and turbulence fluctuations were added downstream of the inlet plane as volume forces. Hanging grid nodes allowed local Cartesian grid refinement for the transport of the inflow turbulence and the capture of the wake dynamics, which extended from inlet to 9 rotor diameters D_{WT} downstream of the WT with a resolution of $\Delta x_{CFD} = 1\,\text{m}$. In addition, the blade tip vortex and rotor area were refined using additional structured meshes, as represented in Figure 5. The grid resolution in the rotor area and blade tip vortex region was $\Delta x_{CFD} = 0.125\,\text{m}$, with intermediate resolutions to ensure a smooth transition to the wake refinement region. The whole computational setup consisted of about 461 Mio. cells.

Table 2. Maritime inflow data.

Airwake Data	Mean Velocity Profile		Mann-Box Parameters	
	V_W, m/s	α	I, %	L, m
C1, C4	11.3	0.14	5.00	45
C2, C5, C7	25.0	0.14	6.84	60

4.1.2. CFD Analysis

Figures 6 and 7 show snapshots of the streamwise velocity distribution and the vortical structures downstream of the WT. In Figure 6, the wake velocity deficit associated with the high induction at this operating point creates a shear layer at the wake boundary. The clearly defined blade tip vortex helix in the near-wake region prevents the turbulent mixing in the wake, which sets on after the blade tip vortices begin to breakdown. In Figure 7, no such distinct shear layer can be distinguished between the wake and the ambient turbulent flow. Less distinct blade tip vortices can be observed, as they breakdown close downstream of the rotor due to interactions with the atmospheric turbulence.

For the following piloted simulations, blade tip vortices from Figure 6 were selected for the simulation of helicopter vortex encounters and are summarized in Table 3. Close, medium and far distances from the WT were chosen to represent vortex encounters at different positions x_{EN} of the flight corridor. The far distance corresponds approximately to the center of the flight corridor (Figure 1a). The selected blade tip vortices were located at the upper boundary of the WT (longitudinal vortex rotor interaction) and at the WT hub height (orthogonal vortex rotor interaction). Note that the blade tip vortices for the wind speed of $V_W = 25.0$ m/s decayed rapidly (Figure 7) and could not be identified over such a far range as in C1. Therefore, the same vortex encountering positions were chosen for C2, and the severity of the stochastic non-directional turbulence was examined.

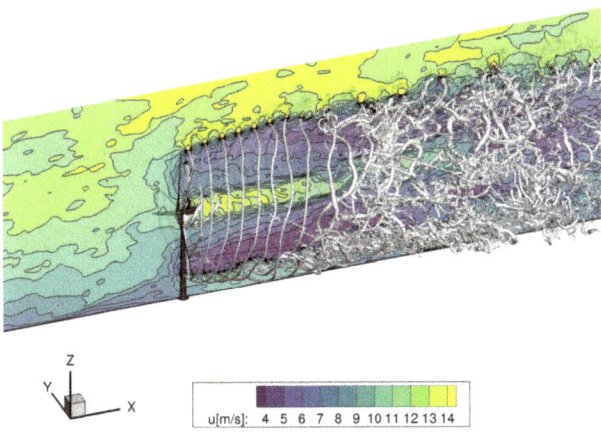

Figure 6. Instantaneous streamwise velocity and λ_2-isosurfaces in the WT wake flow field of C1 at $V_W = 11.3$ m/s. Note that the WT geometry is plotted for more clarity, but was not resolved in the CFD simulation of C1. (OS-1).

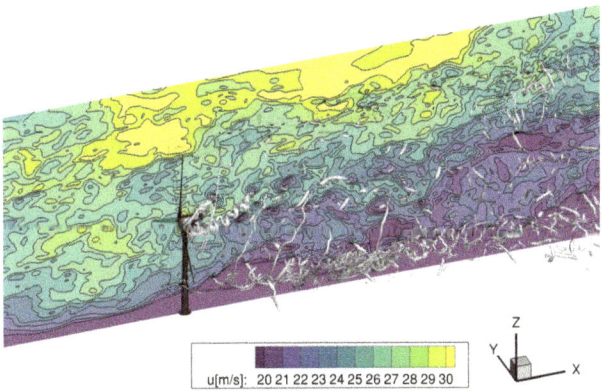

Figure 7. Instantaneous streamwise velocity and λ_2-isosurfaces in the WT wake flow field of C2 at $V_W = 25.0$ m/s. Note that the WT geometry is plotted for more clarity, but was not resolved in the CFD simulation of C2. (OS-1).

Table 3. Selected blade tip vortices for the piloted simulation of vortex encounters with properties identified from the WT wake flow field in C1.

Parameter	Pos. 7	Pos. 5	Pos. 2	Pos. 13	Pos. 11	Pos. 8
	Upper Boundary			Hub Height		
x_{EN}, m	300	175	100	330	150	100
R_c, m	2.96	1.30	1.26	-	0.80	1.00
V_c, m/s	3.20	10.49	9.35	-	8.50	10.57
Γ, m²/s	79	133	116	-	110	98

The blade tip vortex position x_{EN} corresponds to the local maximum of vorticity, and the blade tip vortex core radius R_c is the distance between the core center and the local maximal tangential speed V_c. The blade tip vortex circulation Γ is the integral of the out-of-plane vorticity over a circular area around the vortex center. It shall be mentioned that at position 13 no vortex parameters could be determined, as no clear helical vortex structure could be identified, only an area of globally higher vorticity. In the WT near-wake and in ideal uniform inflow conditions, the strength of the blade tip vortices slowly decayed, and their radii increased as they were convected downstream of the WT, due to diffusivity effects. Here, however, we can observe higher circulation at position 5 than at position 7, which was located further downstream. This was due to the merging of the blade tip vortices at this position, leading to a larger region of high vorticity, a mechanism already reported in [35]. The merging process is visible in Figure 8 through an increase in both the vorticity and the distance between the neighbouring vortex cores.

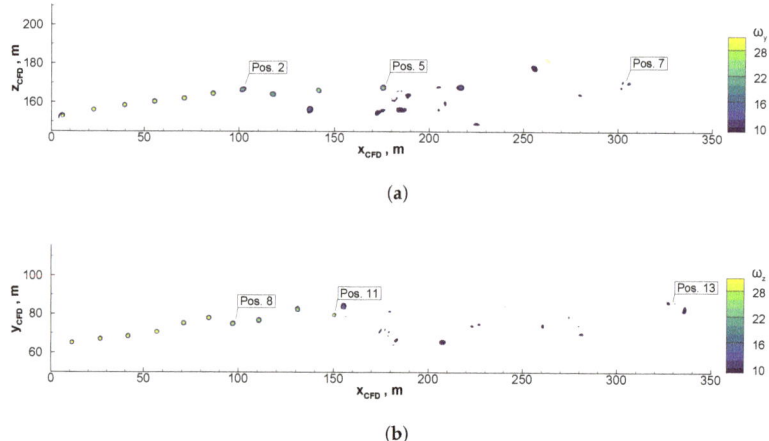

Figure 8. Vortex merging in WT wake flow field C1, (OS-1). (**a**) Vorticity ω_y in a vertical plane passing through the rotor plane at $y_{CFD} = 0$ m. (**b**) Vorticity ω_z in a horizontal plane at $z_{HUB} = 90$ m.

4.2. Non-Rotating WT

In this section, the computational setup and resulting flow fields of OS-2 are presented, which were postprocessed afterwards to airwake data C4, C5 and C7 (Table 1).

4.2.1. CFD Setup

In order to capture the unsteady features of the flow separation in the wake of the nacelle, scale resolving DES methods (Detached-Eddy Simulation) were required. The Menter shear-stress transport (SST) $k - \omega$ model [36] was used for turbulence modeling, and the 5th-order WENO spatial scheme was applied in the wake in order to reduce the numerical dissipation of vortices [26]. For numerical stability reasons, the 2nd-order Jameson–Schmidt–Turkel spatial scheme was used in the boundary-layer-resolving grids around the WT geometry, as mentioned in the previous section. Isotropic local mesh refinement with a cell size of $\Delta x_{CFD} = 0.2$ m was used in the wake of the nacelle, in order to cover all relevant flow dynamics in the flight region of the helicopter. All boundary layers of the WT geometry were resolved so that the dimensionless wall distance satisfied $y^+ \approx 1$. It shall be noted that for the considered WT, the nacelle geometry was approximated as a simple rectangular solid. The whole setup consisted of approximately 115 Mio. cells and is represented on Figure 9. The computational domain extended upstream, laterally, vertically and downstream of the WT, respectively, by 360 m, 350 m, 640 m and 340 m. The same boundary conditions were used at the inlet as for cases C1 and C2; see Table 2. Periodic boundary conditions were used laterally. A far-field boundary condition was used at the top and outlet, and a non-slip condition was imposed at the bottom.

Figure 9. CFD grids around the WT, with the WT surface in red; for more clarity, every 4th grid line is plotted (OS-2).

For the WT wake flow fields C4 and C5, the same inflow data were used as in the OS-1. In order to take into account the influence of another rotating WT located upstream for the WT wake flow field C7, one further simulation at the wind speed of $V_W = 25.0 \, \text{m/s}$ was performed. The upstream WT was simulated with the ACL method with the same turbulent inflow as in the OS-1.

4.2.2. CFD Analysis

In the hover area, the bluff-body aerodynamics of the nacelle with lateral inflow shaped the unsteady flow field characteristics. Figure 10 shows the instantaneous streamwise velocities for the three non-rotating WTs. The recirculation area directly in the shadow of the nacelle appeared in all cases. More important for the helicopter approach maneuver is the vortex shedding that evolved downstream of the nacelle and the boundary layer with a strong velocity gradient and high turbulence intensity developing at the upper surface of the nacelle. The wake dynamics of the nacelle were analyzed in [4,5] along virtual approach paths and compared to a turbulence criterion for safe helicopter operations [37].

Due to its relevance for the following piloted simulation, Table 4 summarizes the average standard deviation of the vertical velocity component σ_w over the five-meter-radius circular area formed by the helicopter rotor above the nacelle at an altitude above ground of $z_{CFD} = 101 \, \text{m}$. In order to capture the effects of longitudinal movements of the helicopter during the hover maneuver, three positions of the center of the rotor area were considered. The ideal position is centered over the nacelle, and two other positions with a downstream shift of 1.5 m and 2.5 m were considered (Figure 11). The 12 s long signal of the WT wake flow field data was used, whereby a high-pass filter at 0.5 Hz and a Hann windowing function were applied before performing the Fast-Fourier Transform to remove transient low-frequency information in the signal.

Figure 11 shows the distribution of the standard deviation of the vertical velocity component σ_w in a plane located at $z_{CFD} = 101 \, \text{m}$. For all cases, vortex shedding downstream of the nacelle is visible by an area of increased values of the standard deviation of the vertical velocity component σ_w. Furthermore, an increase in the standard deviation of the vertical velocity component σ_w upstream of the nacelle is clearly visible between C4 and C5. For the non-isolated WT, only a slight increase in the standard deviation of the vertical velocity component σ_w is visible upstream between C5 and C7.

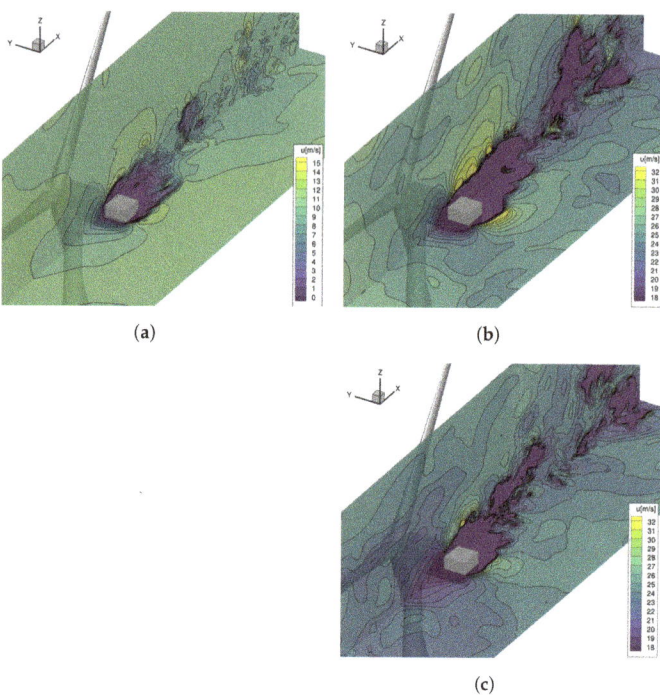

Figure 10. Instantaneous streamwise velocity of WT wake flow field (OS-2). (**a**) C4 − $V_W = 11.3$ m/s; (**b**) C5 − $V_W = 25.0$ m/s; (**c**) C7 − $V_W = 25.0$ m/s.

Table 4. Averaged standard deviation of the vertical velocity component σ_w over the helicopter rotor area from WT flow field data C4, C5 and C7.

Case	Wind Speed V_W, m/s	WT Surrounding	Vertical Standard Deviation		
			σ_{w0}, m/s	$\sigma_{w+1.5}$, m/s	$\sigma_{w+2.5}$, m/s
C4	11.3	isolated	0.42	0.46	0.49
C5	25.0	isolated	0.93	0.99	1.03
C7	25.0	non-isolated	0.99	1.02	1.05

An increase in the wind speed from $V_W = 11.3$ m/s to $V_W = 25.0$ m/s had a negligible influence on the vertical turbulence intensity $I_w = \sigma_w / V_W$ in the helicopter rotor area, as it was located far away enough from the nacelle's turbulent boundary layer. The presence of an upstream WT in contrast led to an increase in the vertical turbulence intensity I_w of more than 5.5% due to the additional turbulence generated by the wake of the upstream WT. In contrast, the standard deviation of the vertical velocity component σ_w in the helicopter rotor area increased strongly with an increase in the wind speed from $V_W = 11.3$ m/s to $V_W = 25.0$ m/s (Table 4). A slight additional increase in the standard deviation of the vertical velocity component σ_w occurred due to the presence of an upstream WT. Furthermore, the standard deviation of the vertical velocity component σ_w increased with a longitudinal shift in the downstream direction.

It was found that the standard deviation of the vertical velocity component σ_w correlated better with the perceived turbulence of the pilots in the following helicopter simulations (Figure 25a), because it is not a normalized measure as the vertical turbulence intensity I_w. The chosen longitudinal shifts of the helicopter rotor area correspond to the performance limits of the following piloted simulations (Table 7). Therefore, the helicopter was affected slightly more by turbulence if its position deviated downstream.

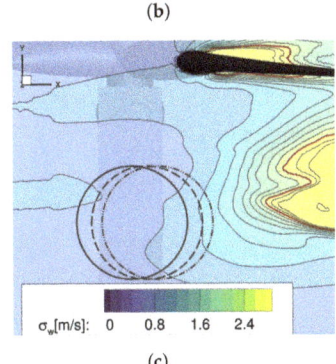

Figure 11. Standard deviation of the vertical velocity component σ_w in a plane located at $z_{CFD} = 101$ m from [4]. The circles represent the integration area of Table 4 (σ_{w0}—solid, $\sigma_{w+1.5}$—dashed, $\sigma_{w+2.5}$—dashed dotted) (OS-2). (a) C4 — $V_W = 11.3$ m/s; (b) C5 — $V_W = 25.0$ m/s; (c) C7 — $V_W = 25.0$ m/s.

5. Piloted Simulation

5.1. Research Flight Simulator AVES

The simulator campaign was conducted at DLR's research flight simulator AVES (Air Vehicle Simulator, Figure 12a [38]), which features a fixed platform and a motion platform with interchangeable cockpits. The piloted simulations were configured to represent DLR's research helicopter ACT/FHS (Active Control Technology/Flying Helicopter Simulator [39]). It is a highly modified version of an EC135 helicopter, which features, among others, a fly-by-light full authority flight control system and is used for flight testing. Its characteristics differ from a standard EC135 helicopter, but it is considered representative of light utility helicopters, which are used in offshore wind farms. For this study, a replica of the ACT/FHS cockpit was used in the AVES motion platform.

A dedicated maritime visual environment with weather effects, dynamic waves and rotating WTs was used (Figure 12b [40]). It provides a realistic cueing environment to the pilot during the simulation campaign, which is important to take into account increased pilot workload due to the lack of available cues while flying offshore missions [41]. The maritime visual environment was configured to represent the offshore wind farm Global Tech I (GTI, Figure 1a), which is located at the German Bight. It consists of 80 WTs with a nominal power of 5 MW, whose properties are similar to the ones of the CFD-simulated WT.

(a) (b)

Figure 12. AVES simulation facility at DLR. (**a**) AVES research simulator. (**b**) Maritime offshore scenario of Global Tech I (GTI).

5.2. Helicopter Modeling

DLR's non-linear real-time helicopter modeling program HeliWorX was used to model the ACT/FHS. It is based on the helicopter modeling program SIMH [42], which was used originally in the former flight simulator to model a Bo105 with a hingeless rotor system, and was adapted to model an EC135 with a bearingless rotor system. The most important properties of the helicopter model can be found in Table 5.

Table 5. Properties of the helicopter model used in the piloted simulation.

Parameter	ACT/FHS
R_H, m	5.1
f_H, 1/min	395
Ω_H, rad/s	41.4
U_H, m/s	211
N_b	4
Mass, kg	2630

Helicopter models in HeliWorX each consist of a set of modular components (fuselage, horizontal stabilizer, vertical stabilizer, main rotor, tail rotor, etc.). The main rotor is modeled as fully articulated with an equivalent hinge offset and spring restraint in order to represent the fundamental flapping and lagging natural frequencies. The main rotor blades are modeled as rigid blades, and 10 blade sections per blade are used to calculate the aerodynamic forces and moments. Furthermore, the dynamic inflow model of Pitt and Peters is used during the piloted simulations [43].

Furthermore, the helicopter model contains additional features, such as an interface for unsteady airwake data to simulate local aerodynamic effects such as wakes from rotating WTs [8]. Unsteady CFD-generated airwake data were superimposed on 43 distributed airload computation points (ACPs, Figure 13) of the helicopter model in total, using spatial and temporal linear interpolation, during the piloted simulation. The unsteady airwake data were looped in time, and temporal blending was performed to enable smooth transition. For the first approach, an interaction between airwake data and tail rotor was implemented.

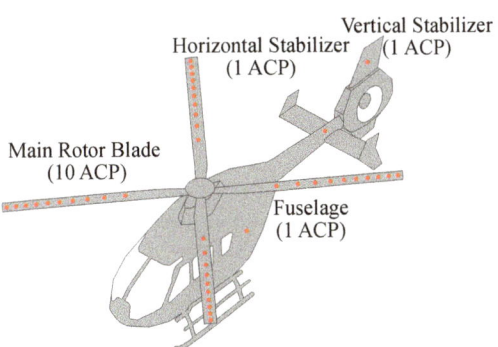

Figure 13. Distribution of airload computation points (ACPs) in HeliWorX.

5.3. Pilot Task

5.3.1. Transit Task

The transit task was performed to examine the size of flight corridor for OS-1 (Figures 1a and 2). It consisted of a straight, level flight with aerodynamic disturbances, delayed pilot response and recovery to the original states. Existing mission task elements and the classification of transients following failures (Figure 8) from ADS-33 [44] were combined for this pilot task. The challenge of OS-1 was the definition of a pilot task with high reproducibility which would lead to vortex encounters between the main rotor hub and the relatively small blade tip vortex cores. It was applied for the selected vortex encounter positions from Table 3.

At the beginning of the simulation, the helicopter was trimmed with an airspeed of $V_\infty = 80$ kt in proximity to the blade tip vortex of the WT, and the main rotor hub encountered the vortex core after $t_{EN} \approx 2$ s without any pilot input. Therefore, the pilot was instructed to avoid any pilot input for the first 2 s of the simulation. After the vortex encounter at t_{EN}, the pilot was instructed to avoid any pilot input for an additional 3 s to simulate a delayed pilot response. Consequently, the pilot started to pull the helicopter back to its original states at $t_{RE} = 5$ s within a stabilization time limit. The start of the recovery maneuver at t_{RE} was indicated by the simulation operator via radio to increase comparability between different simulations. All pilot task performance limits are specified in Table 6.

For this task, a helicopter with bare-airframe response type (BA) and without any stabilization was chosen as a conservative approach to take into account simulation deficiencies. The airspeed of $V_\infty = 80$ kt is rather low for cruise flights of helicopters and corresponds to an advance ratio of $\mu \approx 0.2$. However, this advance ratio still fulfills the guidelines for rigid line vortex modeling approaches of $\mu \geq 0.2$ from [18].

Table 6. Performance limits of pilot task for transit flights (OS-1).

Parameter	Desired	Adequate
Heading, °	±5	±10
Airspeed, kt	±5	±10
Altitude, ft	±10	±30
Stabilize time, s	<5	<8
Response time, s	3	3

Pilots have commented on the task: In reality they would rather gently stabilize the helicopter than fight the turbulence within the wake to recover the original state to avoid pilot-induced oscillations. This would especially be the case for vortex encounters at hub height. Furthermore, the pilots have mentioned the artificially delayed pilot response time is too large to represent reality. The artificially delayed pilot response time was used to increase the effects of the vortex encounter, to enable comparisons between offline analysis and piloted simulations after 3 s and to improve reproducibility. Therefore, the large delayed pilot response time can be considered a conservative approach.

5.3.2. Hover Task

The hover pilot task was defined to examine the lateral safety clearance at the helicopter hoist area for OS-2 (Figures 1b and 3). It combined the hover mission task element from ADS-33 [44] with operational procedures of helicopter hoist operations at offshore WTs. Therefore, the communication between the helicopter pilot and the helicopter hoist operator was simulated as well. The original task consists of an approach, stabilization and precision hovering at the WT wake, but only the hover phase was used for the assessment.

At the beginning of the simulation, the helicopter was trimmed in proximity of the WT hoist position and an oblique approach was performed (Figure 14). After the helicopter was stabilized above the hoist position, precision hovering of 30 s was performed by the pilot. During all phases, the pilot received continuously directions from the hoist operator via radio for positioning the helicopter above the hoist area. The hoist operator was simulated by the simulation operator, who was aware of the exact position of the helicopter via numerical displays. As in operational practice, standardized radio messages to guide the pilot in a horizontal plane with distance information in 1 m steps (e.g., forward-2/right-1) were used.

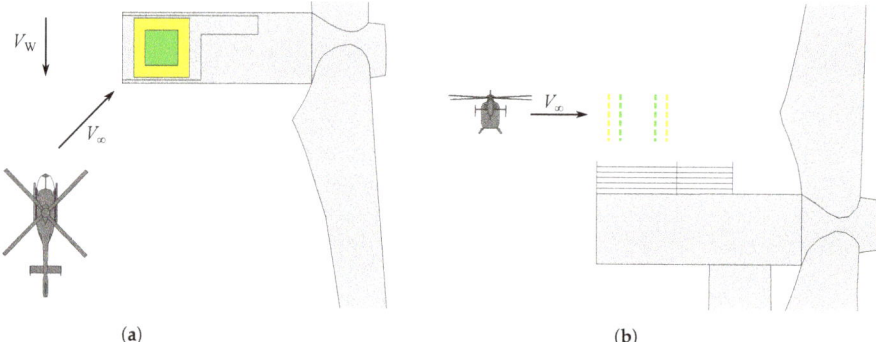

Figure 14. Hover task with performance limits at a non-rotating WT (OS-2). (**a**) Hover task at a non-rotating WT (top view); (**b**) Hover task at a non-rotating WT (side view).

The performance limits of the pilot task can be found in Table 7. The lateral and longitudinal limits were fitted to the size of the WT hoist area and to the step size of the coarse radio guidance. For helicopter hoist operations, gentle altitude corrections are more important for hoist crew comfort than rapid corrections within tight limits. Therefore, all pilots were instructed to hold coarsely to a fixed altitude to increase reproducibility, but no hard altitude performance limits were given. This altitude corresponded to a clearance of 5 m between the helicopter skid and the WT nacelle. Two different helicopter response types of the flight control system were used for the hover task. The helicopter with BA response type corresponds to a helicopter without any stabilization system and was used for a conservative estimate of the lateral safety clearance. In contrast, the helicopter with attitude-command attitude-hold (AC) response type is comparable to the response type of a commercial EC135 helicopter and reduced the workload.

Table 7. Performance limits of the pilot task for hovering (OS-2).

Parameter	Desired	Adequate
Heading, °	±5	±10
Lateral limit, m	±1.5	±2.5
Longitudinal limit, m	±1.5	±2.5
Hover time, s	<30	<30

The pilots said the pilot task well represented reality and was suitable for pilot training, even though no externally slung load was used. As in operational practice, the pilots use the horizontal blade (L-position), the cage of the nacelle and textures on the surface (dirt, rust, etc.) as visual cues to stabilize the helicopter. Therefore, a realistic cueing environment is important, especially for this pilot task. The pilots did suggest slightly loosening the performance limits for the heading to enable a better view of the the WT, because it is not necessary to hold a perfect 90° angle to the WT nacelle, and visual cues were partly hidden by the helicopter cockpit.

5.4. Objective and Subjective Assessments

Objective and subjective assessment methods were used to evaluate the simulation results. The objective methods used the data from the simulations and focused on the helicopter's motion transients and available safety clearance and control margins.

For the offline analysis of OS-1, the requirements for helicopter motion transients from ADS-33 were used (Table 8). The helicopter motion transients were recoverable to a safe steady flight condition without exceptional piloting skill [44]. Originally, the classification was developed to evaluate the motion transients due to a disturbance of the flight control system. The vortex encounter at t_{EN} was considered as a comparable disturbance for "forward flight—near Earth" from Table 8. Therefore, both requirements of "hover and low speed" and "up-and-away" applied. For this study, the assessment of helicopter motion transients was simplified to the requirements of "hover and low speed".

In contrast, the subjective methods were based on pilot rating scales. The pilots were instructed to repeat each test case 2–3 times to familiarize themselves with the new test conditions and to achieve reproducible simulation results. Overall, two experimental test pilots of helicopters and two professional helicopter pilots with different levels of experience participated at the simulation campaign (Table A1). The evaluation of OS-1 was performed by Pilots A, B and D; and the evaluation of OS-2 was performed by Pilots A, B and C. Note that Pilot A and Pilot C used to work as helicopter pilots for German offshore wind farms and are highly familiarized with the examined operational scenarios in reality.

The pilots were asked to evaluate the perceived turbulence at OS-1 and OS-2 with the Turbulent Air Scale (TS, Table A2). It is a subjective measure of the impact of the turbulence on the helicopter.

Furthermore, the Upset Severity Rating (USR, Figure A1) was used to evaluate the severity of the vortex encounter in OS-1. It consists of a decision tree, which originates from pilot rating scales for failure transients and is a subjective measure of the effect of the upset and the ability to recover the helicopter from a disturbance. The usage of the USR-rating was simplified in this study by evaluating solely the ability to recover.

Lastly, the widely used Cooper–Harper Handling Qualities Rating Scale (HQR, Figure A2) was used to evaluate the hover task in OS-2. It consists of a decision tree and is a subjective measure of the additional pilot compensation required to fulfill a specific piloting task. The usage of the HQR ratings is an integral part of experimental test pilot training due to its complexity, and it was only used by Pilots A and B in this study.

Table 8. Requirements for helicopter motion transients following control system failures (ADS-33E-PRF [44]).

Level	Flight Condition		
	Hover and Low Speed	Forward Flight	
		Near Earth	Up-and-Away
1	3° roll, pitch, yaw 0.05 g nx, ny, nz no recovery action for 3 s	both hover and low speed and forward flight up-and-away requirements apply	stay within OFE no recovery action for 10 s
2	10° roll, pitch, yaw 0.20 g nx, ny, nz no recovery action for 3 s	both hover and low speed and forward flight up-and-away requirements apply	stay within OFE no recovery action for 5 s
3	24° roll, pitch, yaw 0.40 g nx, ny, nz no recovery action for 3 s	both hover and low speed and forward flight up-and-away requirements apply	stay within OFE no recovery action for 3 s

6. Results

6.1. Offline Analysis of OS-1

In the following section, the results of the offline analysis of the longitudinal vortex rotor interactions (position 2, position 5, position 7) of OS-1 are shown. Therefore, the vortex encounters of the helicopter at the positions from Table 3 were simulated without any pilot input. The ability of the simulation setup to cause rotor-centered vortex encounters is shown, and the helicopter reactions of the non-piloted simulations were assessed using the ADS-33 offline criteria from Table 8. Furthermore, those results are compared with airwake data from the previously used SWM from [8].

The airwake velocities (V_x, V_y, V_z) of C1 and the SWM were extracted from an ideally straight line through the vortex core at position 2 and are depicted in Figure 15. This straight line corresponds approximately to the flight path trajectory at the beginning of the following helicopter simulations without any pilot. The vortex core center can be identified by the zero crossing of the velocity V_z. Due to the simplifications of the SWM, a perfectly symmetrical velocity profile is shown. In contrast, the velocity profile of C1 shows deformations due to blade tip vortex deformation within the CFD simulation. In addition, C1 contains minor velocity disturbances all throughout due to ambient atmospheric turbulence, which disturbed the flight path even at the beginning of the helicopter simulations and caused deviations from the ideal flight path. Note that the extracted straight line and the helical WT blade tip vortex are inclined (Figure 2). Therefore, Figure 15 is not suitable for determining the vortex core radius.

Figure 16 shows the lateral and vertical deviations ($\Delta x_{\text{non-piloted}}$, $\Delta z_{\text{non-piloted}}$) between the ideally straight line and the non-piloted helicopter simulations of C1 and SWM. The flight direction of the helicopter was west and is indicated by an arrow. Furthermore, the vortex core encounter is indicated at $y_{\text{EN}} = 0$ m. Note that a simulation setup with a perfectly rotor-centered vortex encounter would correspond to zero deviations at y_{EN}. The flight path deviations of the SWM were smaller compared to C1, which amounted to approximately 0.5 m. This was due to the additional ambient atmospheric turbulence in C1 (Figure 15). The vortex core radius at position 2 was estimated to be $R_c = 1.26$ m (Table 3), which is larger than the flight path deviations. Therefore, the vortex core was encountered by the helicopter rotor hub, but the vortex core center was missed.

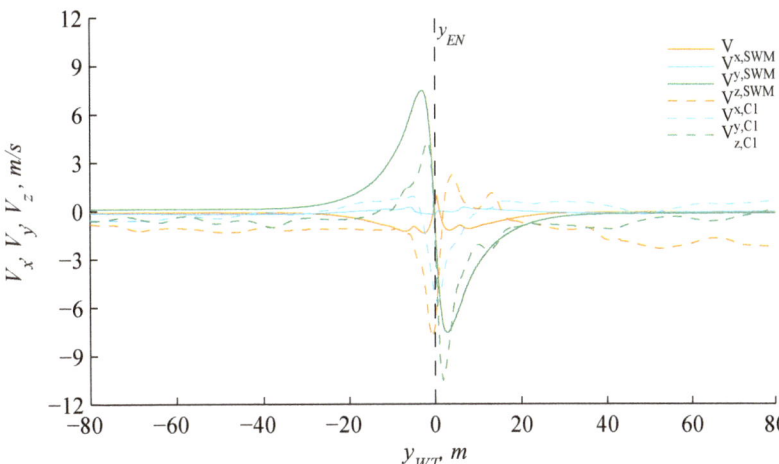

Figure 15. Comparison between the airwake velocities (V_x, V_y, V_z) of C1 and SWM at position 2.

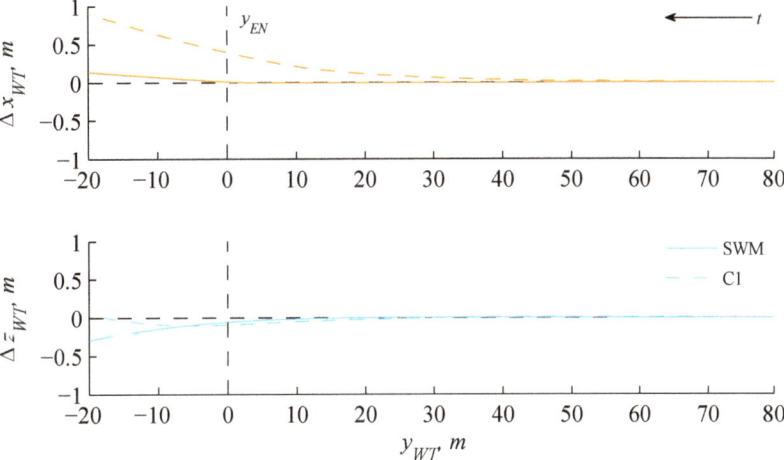

Figure 16. Lateral and vertical flight path deviations ($\Delta x_\text{non-piloted}$, $\Delta z_\text{non-piloted}$) of the non-piloted simulation from the ideally straight line for C1 and SWM at position 2.

Figure 17 shows a comparison of the airwake velocities (V_x, V_y, V_z) between the non-piloted simulation and an extracted ideally straight line for C1 at position 2. Note that a simulation setup with a perfectly rotor-centered vortex encounter would match the curves before the vortex encounter ($y > 0$ m), but deviations between both curves always appeared after the vortex encounter ($y < 0$ m) due to the helicopter's reaction. It can be seen that the airwake velocities (V_x, V_y, V_z) are in a good agreement before the vortex encounter, even though the vortex core center was missed and flight path deviations of approximately 0.5 m appear. This analysis was successfully repeated for position 2, position 5 and position 7 for both flight directions. Thus, the simulation setup is considered acceptable for the simulation of vortex encounters, even though the vortex core was not perfectly hit.

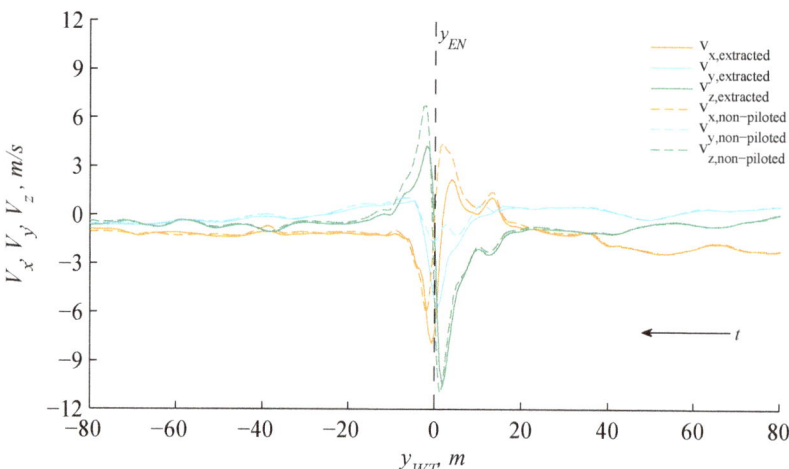

Figure 17. Comparison of the airwake velocities (V_x, V_y, V_z) between the non-piloted simulation and an extracted ideally straight line for C1 at position 2.

Figure 18 shows exemplary attitude changes ($\Delta\Phi$, $\Delta\Theta$, $\Delta\Psi$) and load factors (n_x, n_y, n_z) for the non-piloted simulation of C1 at position 2. The limits of the ADS-33 offline criteria from Table 8 are indicated by horizontally dashed lines. The evaluation time frame for the ADS-33 criteria started at the vortex encounter t_{EN}, lasted for 3 s and ended at t_{OBS}. At the beginning of the simulation, the helicopter was trimmed and a minor drift of the attitudes developed. At $t \approx 1.8$ s, the first effects of the blade tip vortex on the helicopter are visible. After the vortex encounter t_{EN}, the roll $\Delta\Phi$ and pitch $\Delta\Theta$ attitude diverged, and the yaw $\Delta\Psi$ attitude oscillated. The largest load factor was n_z, which appeared directly after the vortex encounter t_{EN}.

It can be seen that the maximum level of the ADS-33 offline criteria was determined by the load factor n_z and the pitch attitude change $\Delta\Theta$. As described in detail in [16], this can be explained by the longitudinal vortex rotor interaction. Due to the aerodynamic excitation of the main rotor by the blade tip vortex, the rotor responded mainly with longitudinal flapping. As a result, large pitch moments of the helicopter occurred, which caused pitch down or up movements, depending on the flight direction through the blade tip vortex. Consequently, the following assessment of the non-piloted simulations could be reduced to the load factor n_z and the pitch attitude change $\Delta\Theta$.

Figure 19 shows a comparison of the pitch attitude change $\Delta\Theta$ and the load factor n_z for the non-piloted simulation of C1 between position 2, position 5 and position 7. The limits of the ADS-33 offline criteria are indicated by horizontally dashed lines, and only non-piloted simulations with flight direction west are depicted. As indicated, the helicopter's reactions strongly depended on the vortex encounter position. Due to the ambient atmospheric turbulence and vortex deformations, the helicopter's reactions deviated from clear pitch-up or pitch-down behavior and even changed its direction. In contrast, the non-piloted simulations of the SWM showed expected, clear pitch-up or pitch-down behavior, depending on the flight direction, though they are not shown here.

A summary of the assessment with the ADS-33 offline criteria is depicted in Table 9. It shows the results of all vortex encounter positions with longitudinal vortex rotor interaction (position 2, position 5, position 7), both flight directions and both airwake data (C1 SWM). If possible, the determining factor ($\Delta\Theta$ or n_z) is indicated by a superscript. It can be seen that the impacts of the vortex encounters were at the border of levels 2–3 and above, independently of the flight direction. The levels of SWM do not show a noticeable dependency on the distance to the WT. In contrast, the levels of C1 increased at position 5 and position

7, which were further away from the WT. Therefore, an increase in the impacts of the vortex encounters can occur with increasing distance to the WT. This can be explained at position 5 with effects such as vortex merging (Figure 8) and vortex deformations, which may cause larger helicopter reactions.

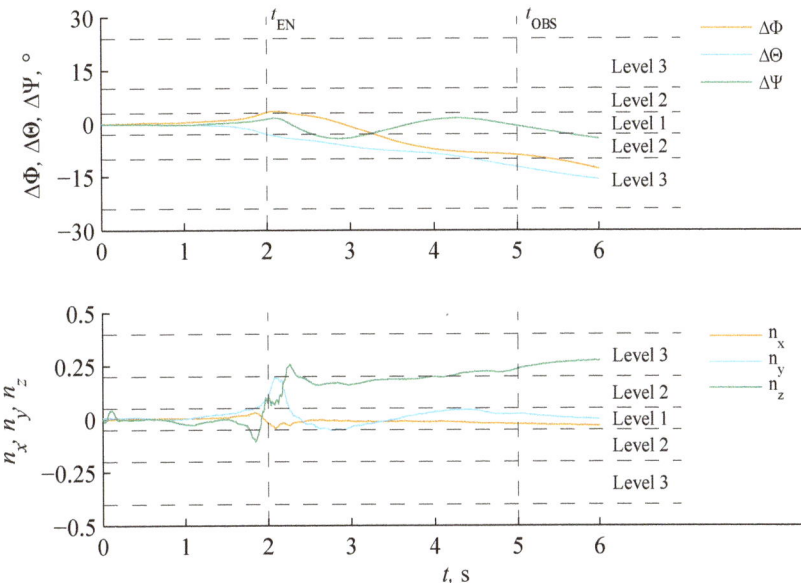

Figure 18. Attitude changes ($\Delta\Phi, \Delta\Theta, \Delta\Psi$) and load factors ($n_x, n_y, n_z$) for the non-piloted simulation of C1 at position 2 with limits of ADS-33 offline criteria.

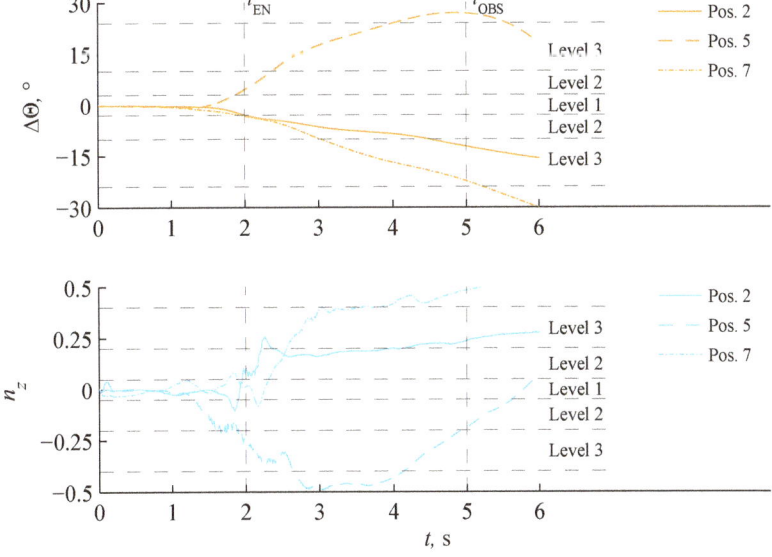

Figure 19. Comparison of pitch the attitude change $\Delta\Theta$ and the load factor n_z for the non-piloted simulation of C1 between position 2, position 5 and position 7 with limits of ADS-33 offline criteria (flight direction west).

Table 9. Summary of the assessment of the non-piloted simulations with the ADS-33 offline criteria.

Flight Direction	Airwake Data	Pos. 2 $x=100\,\text{m}$	Pos. 5 $x=175\,\text{m}$	Pos. 7 $x=300\,\text{m}$
West	C1	2–3	>3	>3 [+]
	SWM	2–3	2–3	2–3
East	C1	2–3 [+]	>3	3 [+]
	SWM	2–3	2–3	2–3

[+]: Loadfactor.

According to the ADS-33 offline criteria in Table 8, vortex encounters larger than level 3 may occur at medium to far distances from the WT, which violates the specifications. These predictions are further examined with piloted simulations in the next section.

6.2. Piloted Simulation

6.2.1. Results of OS-1

For the piloted simulations of OS-1, all vortex encounter positions from Table 3 were examined with airwake data C1 and C2. Due to the limited time of the campaign, only transit flights with flight direction west were examined. Exemplary results are given for position 5, which had the largest predicted vortex impact in the offline analysis (Table 9).

Figure 20a,b shows exemplary the pilot input (δ_{lat}, δ_{lon}, δ_{ped}, δ_{col}) and the helicopter reactions (Φ, Θ, Ψ, ΔH) for position 5 at the upper boundary (Figure 2). As already described, ambient atmospheric turbulence caused a small drift at the beginning of the simulation. At $t \approx 1.7\,\text{s}$, the first influence of the WT blade tip vortex can be seen by a pitch-up movement. This pitch-up movement increased and the helicopter started to gain altitude ΔH, even after the vortex core was encountered at $t_{EN} \approx 2.0\,\text{s}$. An unstable oscillation occurs in roll Φ and yaw Ψ axes. The pilot was allowed to start the recovery maneuver at $t_{RE} = 5.0\,\text{s}$, but the actual recovery maneuver started slightly delayed at $t = 5.6\,\text{s}$. This was due to delays between radio instructions by the simulation operator and the actual response time of the pilot. Mainly pitch δ_{lon} and roll inputs δ_{lat} were given by the pilot to recover the helicopter. Very low activity on pedal input δ_{ped} is used to stabilize the yaw axis Ψ and no collective inputs δ_{col} were used to correct the altitude. The helicopter was recovered and stabilized after $t = 14.0\,\text{s}$.

Overall, relative attitude changes amounted to 20° for the pitch axis Θ and to $\pm 15°$ for roll Φ and yaw Ψ axes. Furthermore, an altitude gain ΔH of up to 30 m occurred. The largest pilot inputs were pitch δ_{lon} and roll inputs δ_{lat} with a range of $\pm 15\%$ relative to the trim with sufficient control margins.

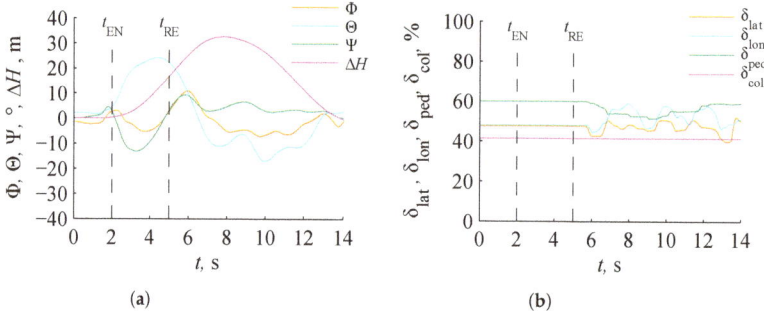

Figure 20. Helicopter reactions and pilot input for the transit task (Pilot A, position 5, $V_W = 11.3\,\text{m/s}$). (a) Attitude (Φ, Θ, Ψ) and altitude (ΔH). (b) Pilot input (δ_{lat}, δ_{lon}, δ_{ped}, δ_{col}).

Figure 21 shows a comparison of the flight path deviations ($\Delta x_{piloted}$, $\Delta z_{piloted}$) and airwake velocities (V_x, V_y, V_z) between the piloted simulation and an extracted ideal, straight-line flight path. The helicopter was trimmed with minor deviations from the ideal flight path, started to drift slightly at the beginning of the simulation and encountered the vortex core at $y_{EN} = 0.0$ m with a deviation of approximately 0.5 m (Figure 21a). For the longitudinal vortex rotor interaction, the airwake velocity V_z is most relevant. Figure 21b shows that the velocity profile and the peak value were sufficiently well captured till the vortex encounter at $y_{EN} = 0.0$ m, and the subsequent deviations were based on helicopter reactions. Note that Figure 21b shows an oblique vortex encounter and is not suitable for determining the vortex core radius R_c.

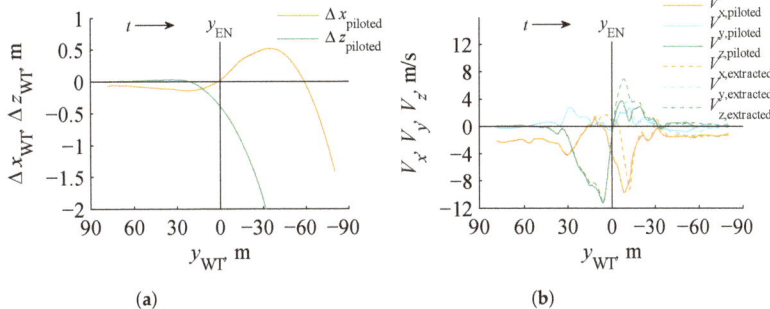

Figure 21. Comparison of flight path and airwake velocities between piloted simulations and an extracted ideal, straight-line flight path (Pilot A, position 5, $V_W = 11.3$ m/s). (**a**) Flight path deviations ($\Delta x_{piloted}$, $\Delta z_{piloted}$). (**b**) Airwake velocities (V_x, V_y, V_z).

Figure 21b shows an unusually wide area (10 m $< y <$ 30 m) with high values of the airwake velocity V_z. As shown in Figure 17, rather fine peak values of the airwake velocity V_z were expected from C1. This can be explained by deformations of the blade tip vortex helix within the CFD simulation. Figure 22 shows the blade tip vortex helix at position 5 and the extracted ideal, straight-line flight path. The ideal flight path is nearly parallel to the vortex axis at the upper boundary. Therefore, the influence of the vortex encounter on the helicopter was prolonged for this flight path, as indicated, which caused larger helicopter deviations. Furthermore, additional aerodynamic effects such as vortex merging were found for position 5 (Figure 8).

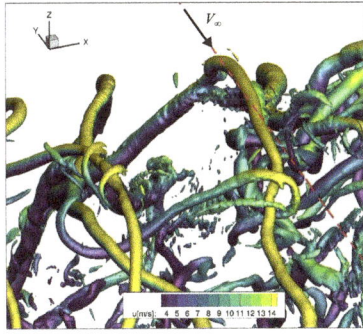

Figure 22. Visualization of the blade tip vortex helix at $V_W = 11.3$ m/s and ideal flight path (red dashed) at position 5 (CFD coordinate system).

Figure 23 shows the subjective pilot assessments of OS-1. The ratings of the three pilots are split into the airwake data C1 ($V_W = 11.3$ m/s) and C2 ($V_W = 25.0$ m/s) and different flight altitudes (Up—altitude upper boundary, Hub—altitude hub height). Each case is described by minimum, maximum and average rating. Note that the position order at hub height (position 7, position 5, position 2) and at the upper boundary (position 13, position 11, position 8) represent increasing distance to the WT from right to left.

The TS ratings in Figure 23a partly contain a high scatter and a range of up to TS-6 (moderate turbulence). For transit flights at hub height, a weak trend to larger perceived turbulence in proximity of the WT might be visible for both wind speeds V_W, but must be considered with care due to the high scatter. In contrast, there is no clear trend for transit flights at the upper boundary, but position 5 at $V_W = 11.3\,\text{m/s}$ and position 2 at $V_W = 25.0\,\text{m/s}$ are of particular interest. As already described, the former caused larger perceived turbulence due to the aerodynamic effects. The latter suffered from unfavorable helicopter trim conditions, which caused large drifts at the beginning of the simulation.

The scatter of the USR-ratings in Figure 23b is lower compared to the TS ratings, and all USR-ratings are below D (minor hazard). A potential trend toward increased ratings in proximity of the WT is visible for transit flights at hub height for both wind speeds V_W. However, there is still no trend visible for transit flights at the upper boundary. Once more, position 5 at $V_W = 11.3\,\text{m/s}$ and position 2 at $V_W = 25.0\,\text{m/s}$ are conspicuous.

Both subjective pilot assessments show similarities. However, the pilots were more confident evaluating the vortex encounter with the USR scale, which can be seen by the decreased scatter. Overall, the pilots perceived moderate turbulence and somewhat large helicopter reactions due to the artificial pilot response time, but the recovery of the helicopter was never considered a major hazard. Consequently, the impact of the same vortex encounter was considered less severe in the piloted simulations compared to the offline analysis. These results were achieved by very conservative assumptions (artificial pilot response time, helicopter with BA response type, rotor-centered vortex encounters) and with pilots with very different flight experience.

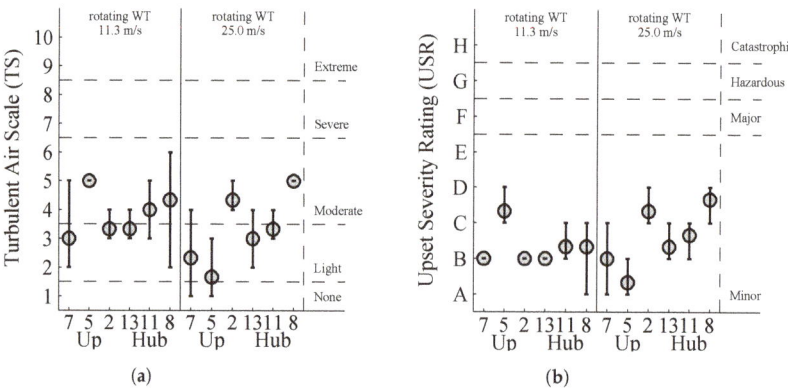

Figure 23. Subjective pilot assessment of the transit task at different positions (Up—altitude upper boundary, Hub—altitude hub height). (**a**) Turbulent Air Scale (TS, Table A2); (**b**) Upset Severity Rating (USR, Figure A1).

6.2.2. Results of OS-2

For the piloted simulations of OS-2, the hover task was performed with airwake data C4, C5 and C7 (Table 1). Figure 24 shows exemplary the longitudinal Δx and lateral Δy positions of the helicopter winch above the WT nacelle of Pilot A for different wind conditions and different helicopter response types (BA, AC). Furthermore, the longitudinal and lateral limits from Table 7 are depicted, and only the 30 s hover phase is shown.

For the helicopter with BA response type, desired task performance was only achieved at a medium wind speed of $V_W = 11.3\,\text{m/s}$ (Figure 24a). The task performance decreased to adequate with a wind speed of $V_W = 25.0\,\text{m/s}$. The additional turbulence for the non-isolated WT caused a further decrease in task performance, but it still remained adequate. Figure 24b shows that in general the task performance improved for the helicopter with AC response type. The fluctuations decreased, and both cases with isolated WTs clearly allowed the desired task performance. For the non-isolated WT, no task performance improvements and lateral deviations of up to $\Delta y \approx +2.5$ m are visible, which correspond to a drift towards the WT rotor disk. It must be noted that Pilot A showed a loss of concentration and exhaustion during the last simulations. It is considered that the strong decrease in pilot performance for the non-isolated WT was based on human factors rather than on turbulence.

Overall, the largest lateral deviations of $\Delta y_{max} = \pm 3.0$ m were detected for the helicopter with BA response type by another pilot and are not displayed here. Therefore, this value is considered as a conservative estimate for potentially critical drifts towards the WT rotor disk and corresponds to a usage of 60% of the lateral safety clearance. Note that for the EC135, a lateral safety clearance of ≈ 5 m was applied (Figure 1b).

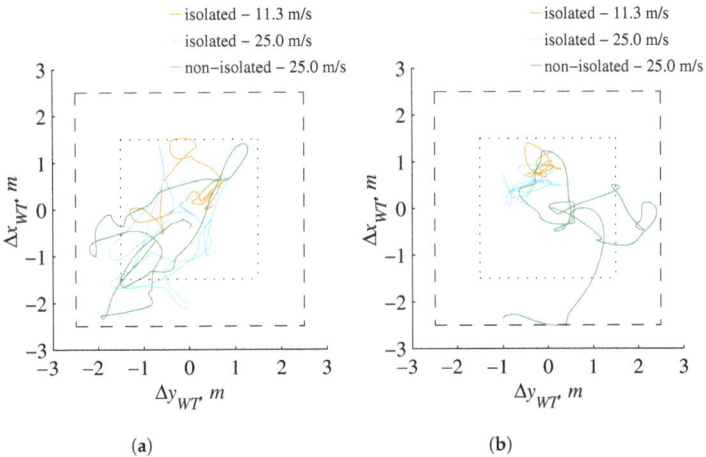

Figure 24. Longitudinal Δx and lateral Δy position of the helicopter winch above the WT nacelle (Pilot A). (**a**) Helicopter with bare-airframe response type (BA). (**b**) Helicopter with attitude-command attitude-hold response type (AC).

Figure 25 shows the subjective pilot assessment of the hover task. The ratings of the pilots are split into the different wind conditions and helicopter response types. The results of the TS rating are described by minimum, maximum and average rating (Figure 25a). In contrast, the HQR ratings have only been evaluated by the experimental test pilots and are depicted separately (Figure 25b).

The TS ratings in Figure 25a have a high scatter and range up to TS-7 (severe turbulence). A comparison between the helicopter response types shows that the perceived turbulence decreased in any one wind condition for the helicopter with AC response type. Consequently, deficiencies of a poor helicopter response type were perceived wrongly as additional turbulence. Furthermore, a clear increase in the perceived turbulence is visible for an increase in the wind speed to $V_W = 25.0$ m/s. For the case of the non-isolated WT, only a slight increase in the perceived turbulence is visible. This behavior fits the standard deviation of the vertical velocity component σ_w in the helicopter rotor area of Table 4.

The most HQR ratings are mostly within level 2, which is considered acceptable for this pilot task (Figure 25b). Only two single ratings of the helicopter with BA response type are within level 3 and are not considered acceptable. The ratings of the helicopter with AC response type are lower or equal compared to the helicopter with BA response type. Furthermore, an increase in the wind speed to $V_W = 25.0$ m/s increased the HQR ratings. However, the HQR ratings did not further increase for the non-isolated WT, even though slightly higher turbulence was perceived. In general, the HQR ratings of Pilot B were higher compared to Pilot A. It is assumed that this behavior was based on the practical experience of Pilot A in offshore wind farms (Table A1). The HQR ratings show that handling qualities of the helicopter with AC response type are still acceptable for this pilot task, but improvements are recommended.

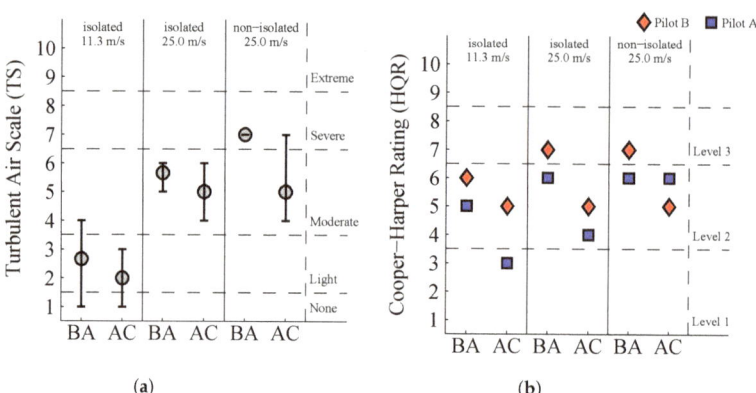

Figure 25. Subjective pilot assessment of the hover task at different positions (BA—helicopter with bare-airframe response type, AC—helicopter with attitude-command attitude-hold response type). (**a**) Turbulent Air Scale (TS, Table A2). (**b**) Cooper–Harper Handling Qualities Rating Scale (HQR, Figure A2).

7. Discussion

7.1. Discussion of OS-1

Offline analysis and piloted simulations in OS-1 were used to examine the size of the flight corridor within an offshore wind farm. The definition of a pilot task with high reproducibility was a challenge, because it is difficult to hit the relatively small vortex cores with the helicopter's main rotor in a comparable way. Trade-offs were made between the depth of the simulation for the pilots (very short simulation time before vortex encounter $t_{EN} \approx 2$ s), precision of vortex encounter (deviations $\Delta x_{piloted}$ and $\Delta z_{piloted}$) and degree of realism (artificial pilot response time of 3 s). The resulting pilot task is considered suitable with sufficient precision of vortex encounters (deviations $\Delta x_{piloted}$ and $\Delta z_{piloted}$ of approximately ± 0.5 m) compared to the vortex core radius (0.80 m $< R_c <$ 2.96 m).

Simulation improvements might be made by an automatic flight path control system which holds a fixed trajectory through the vortex core. After the vortex encounter, the flight path control system should automatically be turned off and the helicopter recovered manually by the pilot. The usage of an artificial pilot response time is recommended to simulate divided attention of the pilot. The artificial pilot response time of 3 s was chosen for direct comparison with the ADS-33 offline criteria (Table 8) and to increase the partly weak effects of the vortex encounter for the analysis. For less conservative and more practical approaches, a reduced artificial pilot response time is recommended. Furthermore, the flight direction of the helicopter should be aligned with the local vortex axis at the vortex encounter to maximize its effect. Lastly, the influence of WT wakes of more than 5 MW only were approximated in previous studies [8]. In consideration of the trend toward more powerful offshore WTs, those examinations should be verified with flow fields from high-fidelity CFD computations.

The transit flight results were achieved by conservative assumptions (long artificial pilot response time, helicopter with BA response type, rotor-centered vortex encounters) and with pilots with very different flight experience. Furthermore, the simulations were partly performed very close to the WT (operational practice ≈ 330 m in GTI), which tends to increase the impact of the vortex encounter due to vortex aging effects [14]. The rather low airspeed of $V_\infty = 80$ kt was expected to increase the impact of the vortex encounter, because the helicopter would remain longer in the influential area of the blade tip vortex. This corresponds to an advance ratio of $\mu \approx 0.2$, which fulfills the guidelines for rigid line vortex modeling approaches from [18]. Effects which decreased the impact of the vortex encounter were neglected (vortex deflection [17], stabilization through a flight control system, etc.).

Overall, the results are considered conservative estimates of the impacts of vortex encounters at current operational offshore wind farms. Nevertheless, additional validation by measurements of WT wake and flight tests with instrumented helicopters in offshore wind farms are recommended.

7.2. Discussion of OS-2

Piloted simulations of the hover task in OS-2 were used to examine the lateral safety clearance to the helicopter hoist area of offshore WTs. In reality, helicopter hoist operations are complex maneuvers, consisting of precision hovering, an additional slung load and communication between the pilot and the hoist operator in an adverse environment. In this study, it was approximated by a hovering task with simulated communication between the pilot and a hoist operator. The main focus of this study was to investigate the influence of turbulence on the helicopter in a realistic simulation. A common medium wind speed of $V_W = 11.3$ m/s and a very high wind speed of $V_W = 25.0$ m/s, which is slightly above usual operating procedures, was chosen. Furthermore, the impact of additional turbulence from a rotating WT upstream of the helicopter hoist area of the non-rotating WT at a wind speed of $V_W = 25.0$ m/s was examined. The same case was considered for a wind speed of $V_W = 11.3$ m/s, at which the upstream rotating WT operated with the expected strong blade tip vortices, but this was not examined due to limited time.

A possible improvement to the simulation would be the implementation of a slung load in the pilot task, which should be affected by the turbulence. Note that in reality the hoist operator would dampen the slung load's movements by horizontally pushing and pulling the hoist cable, which cannot be directly simulated in a flight simulator. Further aspects such as bad weather and low visibility would have an influence on the task performance as well, and should be examined in future studies.

Nonetheless, the results of the hover flight were achieved by conservative assumptions (helicopter with BA response type without stabilization, unusually high wind speed, additional turbulence from an upstream WT) and with pilots with very different flight experience. In addition, piloted simulations for comparisons were performed with a helicopter with AC response type, whose characteristics are comparable to the response type of a commercial EC135 helicopter.

Overall, the results are considered conservative estimates of helicopter hoist operations at offshore WTs, and the simulation fidelity received favorable comments from the pilots. However, a validation of the simulation results by flight tests with instrumented helicopters in offshore wind farms is recommended.

8. Conclusions

A simulation campaign using a light utility helicopter at AVES was conducted to assess potential risks in offshore wind farms for helicopter operations. Two operational scenarios were derived through consultation with the German authorities LBA and BSH to examine transit flights within offshore wind farms and hover flights at single offshore WTs. Both were assessed in a dedicated maritime visual environment of the wind farm GTI with weather effects, dynamic waves, rotating WTs and turbulence interactions based on highly resolved CFD-based flow fields. Conservative assumptions were made to take into account simulation deficiencies and to estimate operational limits. Based on this effort, the following key conclusions were made:

- Transit flight
  - The helicopter's reactions with CFD flow fields can be larger than in the SWM due to blade tip vortex helix deformation, vortex merging and additional turbulence. Those aerodynamic effects can increase the impacts of vortex encounters at medium to far distances from the WT.

- Non-piloted simulations of longitudinal vortex rotor interaction between a helicopter with BA response type and a WT blade tip vortex helix may cause vortex encounters higher than level 3 by ADS-33 offline criteria.
- In contrast, piloted simulations with an artificial pilot response time of 3 s showed that those helicopter reactions can be recovered with little danger.
- Overall, the simulations suggest that the sizes of current flight corridors in offshore wind farms are sufficiently large for the considered scenario. Transit flights at different altitudes, in close proximity to WTs and at various wind speeds, have always been recovered without much risk.

• Hover flight
- The perceived turbulence and the pilot compensation increased with increasing wind speed V_W. Additional turbulence at a non-isolated WT was perceived, but for this specific case it did not necessarily cause additional pilot compensation.
- The largest lateral deviations of $\Delta y_{max} = \pm 3.0$ m were estimated with a helicopter with BA response type and at an unusually high wind speed of $V_W = 25.0$ m/s. Consequently, the lateral safety clearance towards the WT was made 60%.
- Overall, the simulations suggest that the lateral safety clearance is sufficiently large for the considered scenario. Probably, hoist crew comfort and safety are more limiting than the lateral safety clearance.

Future work should focus on validation of the simulation results and the consideration of more powerful offshore WTs. For the CFD computations, additional flow field measurements in the wake of offshore WTs are desirable to evaluate the blade tip vortex aging behind large-scale WTs. In regard to the piloted simulations, further quantitative validation is needed for a profound transfer of the simulation results to reality. Therefore, flight tests with instrumented helicopters in offshore wind farms are desired. Afterwards, a validated simulation with potential improvements could be used to develop recommendation for dimensioning of safety clearances.

Author Contributions: Formal analysis, A.Š., D.H.G. and M.C.; investigation, F.H.; writing—original draft preparation, A.Š., D.H.G. and M.C.; writing—review and editing, A.Š. and T.L.; supervision, F.H., A.Š. and T.L. All authors have read and agreed to the published version of the manuscript.

Funding: This research was funded by the German Federal Ministry of Economic Affairs and Energy (BMWi), grant numbers 0324121A, 0324121B, 0324121C and 0324121D.

Supported by:

Federal Ministry
for Economic Affairs
and Energy

on the basis of a decision
by the German Bundestag

Institutional Review Board Statement: Not applicable.

Informed Consent Statement: Informed consent was obtained from all subjects involved in the study.

Data Availability Statement: Not applicable.

Acknowledgments: The authors would like to thank BSH, LBA, the Flight Service of the German Federal Police (BPOLFLS), Helicopter Travel Munich GmbH (HTM) and Windpark Heliflight Consulting GmbH (WHC) for their professional support and involvement, especially during the research work concerning regulations and operational procedures for helicopter operations in offshore wind farms. Furthermore, the authors would like to thank the helicopter pilots of LBA and BPOLFLS for the participation in the simulation campaign at AVES. Lastly, the authors gratefully acknowledge the High-Performance Computing Center Stuttgart (HLRS) for providing the computational resources necessary for the CFD simulations within the project WEALoads.

Conflicts of Interest: The authors declare no conflict of interest.

Nomenclature

The following nomenclature is used in this manuscript:

D_H	m	Rotor diameter of helicopter rotor
D_{WT}	m	Rotor diameter of WT
f_H	1/min	Rotational frequency of helicopter rotor
f_{WT}	1/min	Rotational frequency of WT
ΔH	m	Altitude change
I	%	Turbulence intensity
I_w	%	Turbulence intensity, vertical direction
L	m	Integral length scale
L_{WT}	m	Distance between WTs in OS-2
$Mass$	kg	Mass of helicopter
N_b	-	Number of blades of helicopter rotor
n_x, n_y, n_z	-	Load factors
R_c	m	Blade tip vortex core radius
R_H	m	Rotor radius of helicopter rotor
t_{EN}	s	Time of vortex encounter
t_{OBS}	s	Time of the end of offline evaluation
t_{RE}	s	Time of the start of recovery maneuver
Δt_{WT}	s	Temporal discretization of airwake data
T_{WT}	s	Temporal dimension of airwake data
u, v, w	m/s	CFD flow fields velocities
U_H	m/s	Blade tip speed of helicopter rotor
V_c	m/s	Blade tip vortex tangential speed
V_W	m/s	Wind speed
$V_{W,\Omega max}$	m/s	Rated wind speed of WT
$V_{W,max}$	m/s	Cut-off wind speed of WT
V_x, V_y, V_z	m/s	Airwake velocities
V_∞	kt	Airspeed of helicopter
$x_{CFD}, y_{CFD}, z_{CFD}$	m	CFD coordinate system
x_{WT}, y_{WT}, z_{WT}	m	WT coordinate system
x_{EN}, y_{EN}, z_{EN}	m	Position of Vortex encounter
$\Delta x, \Delta y, \Delta z$	m	Helicopter position deviations
Δx_{CFD}	m	Local spatial discretization of CFD data
$\Delta x_{WT}, \Delta y_{WT}, \Delta z_{WT}$	m	Spatial discretization of airwake data
X_{WT}, Y_{WT}, Z_{WT}	m	Spatial dimension of airwake data
y^+	-	Dimensionless wall distance
z_{HUB}	m	WT hub height
α	m²/s	Wind profile power law with the exponent
Γ	m²/s	Blade tip vortex circulation
Γ_0	m²/s	Initial blade tip vortex circulation
$\delta_{lat}, \delta_{lon}, \delta_{ped}, \delta_{col}$	%	lateral, longitudinal, pedal and collective pilot input
ϵ	-	Regularization kernel
μ	-	Helicopter advance ratio
σ_w	m/s	Standard deviation, vertical direction
Φ, Θ, Ψ	°	Roll, pitch and yaw attitude
ω_y, ω_z	1/s	Vorticity
Ω_H	rad/s	Rotor rotational speed of helicopter rotor

Appendix A

Table A1. Overview of pilot experience.

	Pilot A	Pilot B	Pilot C	Pilot D
Pilot license	27 years	41 years	20 years	6 years
Experimental test pilot	Yes	Yes	No	No
Aircraft experience: EC135	400 h	1250 h	2045 h	600 h
Aircraft experience: Bo105	200 h	3050 h	-	-
Aircraft experience: Sea King	2500 h	-	-	-
Aircraft experience: Chinook	500 h	-	-	-
Aircraft experience: Bell 205	-	500	-	-
Aircraft experience: Bell 412	-	500	-	-
Aircraft experience: Bell UH-1	-	-	1105 h	-
Aircraft experience: Alouette II	-	1400	138 h	-
Aircraft experience: Agusta A109	-	-	54 h	-
Aircraft experience: Sikorsky S-76	-	-	620 h	-
Aircraft experience: Others	1000 h	-	-	150 h
Total flight hours	4600 h	6700 h	3962 h	750 h
Offshore flights per year	100	1	208	30
Helicopter offshore experience (% of flight hours)	50–75%	0–25%	75–100%	25–50%
Maneuver: Landing OSS	more than 30	0	more than 30	0
Maneuver: Hoisting with person at OSS	more than 30	0	more than 30	0
Maneuver: Hoisting without person at OSS	more than 30	0	0	0
Maneuver: Hoisting with person at ship	0	0	more than 30	0
Maneuver: Ship deck landing	10–30	0	more than 30	0–10

Table A2. Turbulent Air Scale (TS) with turbulence categories none (TS-1), light (TS-2–TS-3), moderate (TS-4–TS-6), severe (TS-7–TS-8) and extreme (TS-9–TS-10) [45].

Scale	Definition	Air Condition
1	-	Flat calm
2	Light	Fairly smooth, occasional gentle displacement
3		Small movements requiring correction if in manual control
4		Continuous small bumps
5	Moderate	Continuous medium bumps
6		Medium bumps with occasional heavy ones
7	Severe	Continuous heavy bumps
8		Occasional negative "g"
9	Extreme	Rotorcraft difficult to control
10		Rotorcraft lifted bodily several hundreds of feet

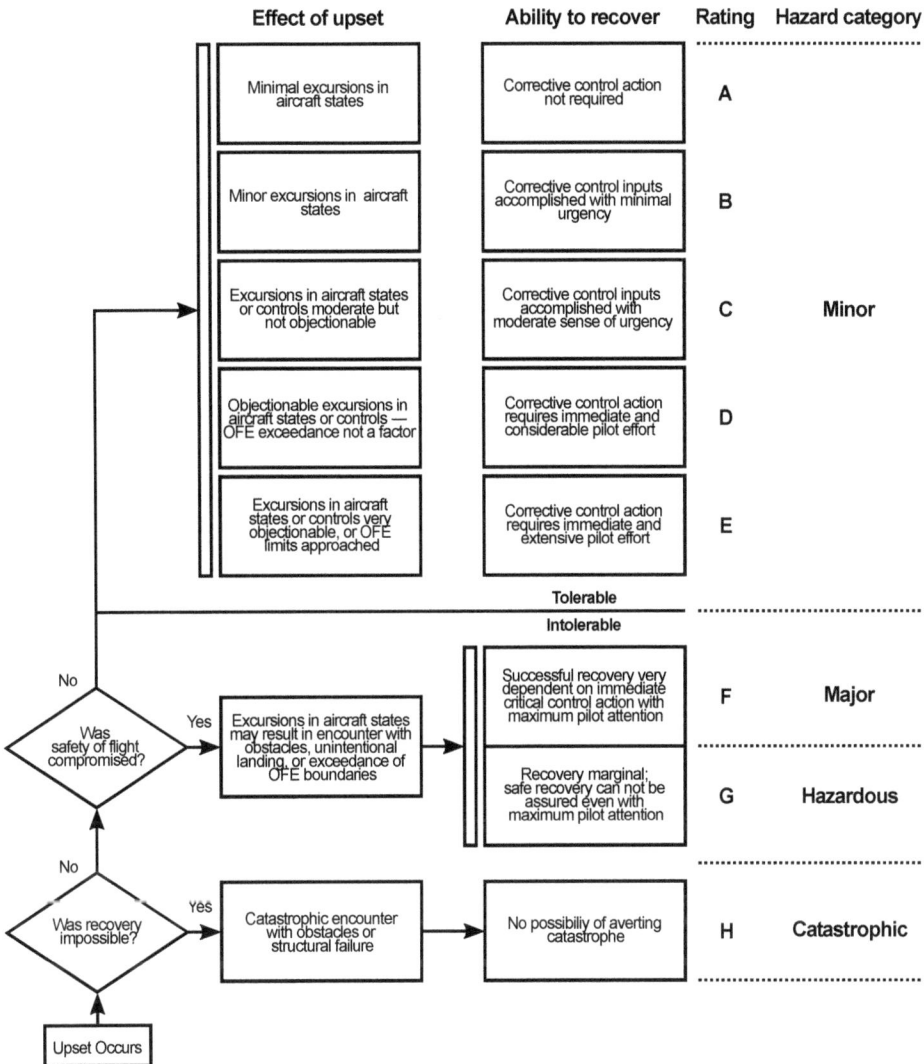

Figure A1. Upset Severity Rating (USR) with hazard categories minor (**A–E**), major (**F**), hazardous (**G**) and catastrophic (**H**) [13].

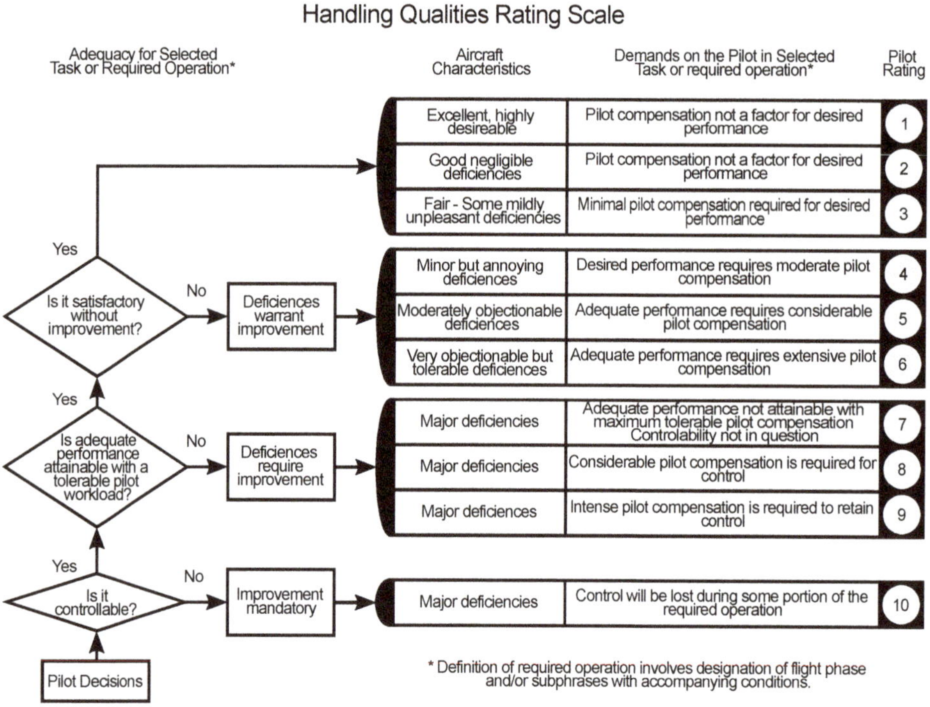

Figure A2. Cooper–Harper Handling Qualities Rating Scale (HQR) with level 1 HQ (HQR-1–HQR-3), level 2 HQ (HQR-4–HQR-6) and level 3 HQ (HQR-7–HQR-8) [46].

Table A3. Douglas sea state scale from World Meteorological Organization (WMO) [47].

Sea State Code	Description of Sea	Significant Wave Height		Wind Speed
		m	ft	kt
0	Calm (Glassy)	0	0	0–3
1	Calm (Rippled)	0 to 0.1	0 to 1/3	4–6
2	Smooth (Wavelets)	0.1 to 0.5	1/3 to 1 2/3	7–10
3	Slight	0.5 to 1.25	1 2/3 to 4	11–16
4	Moderate	1.25 to 2.5	4 to 8	17–21
5	Rough	2.5 to 4	8 to 13	22–27
6	Very Rough	4 to 6	13 to 20	28–47
7	High	6 to 9	20 to 30	48–55
8	Very High	9 to 14	30 to 45	56–63
9	Phenomenal	Over 14	Over 45	64–118

References

1. Mauz, M.; Rautenberg, A.; Platis, A.; Cormier, M.; Bange, J. First identification and quantification of detached-tip vortices behind a wind energy converter using fixed-wing unmanned aircraft system. *Wind Energy Sci.* **2019**, *4*, 451–463. [CrossRef]
2. Bühler, M.; Weihing, P.; Klein, L.; Lutz, T.; Krämer, E. Actuator line method simulations for the analysis of wind turbine wakes acting on helicopters. *J. Phys. Conf. Ser.* **2018**, *1037*, 062004. [CrossRef]
3. Cormier, M.; Bühler, M.; Mauz, M.; Lutz, T.; Bange, J.; Krämer, E. CFD Prediction of Tip Vortex Aging in the Wake of a Multi-MW Wind Turbine. *J. Phys. Conf. Ser.* **2020**, *1618*, 062029. [CrossRef]
4. Cormier, M.; Lutz, T. Numerical Investigation of the Unsteady Flow Field Past an Offshore Wind Turbine in Maintenance Operations. In *STAB/DGLR Symposium*; Springer: Cham, Switzerland, 2020; pp. 592–603.
5. Cormier, M.; Letzgus, P.; Lutz, T.; Krämer, E. CFD Study of an Offshore Wind Turbine in Maintenance Conditions. In *High Performance Computing in Science and Engineering'20*; Springer: Cham, Switzerland, 2021; pp. 435–449.

6. Horvat, B.; Hajek, M.; Rauleder, J. Analysing rotorcraft vortex encounter methods with a lattice-boltzmann method based gpu framework. In Proceedings of the AIAA Science and Technology Forum and Exposition (SciTech), Orlando, FL, USA, 6–10 January 2020.
7. Horvat, B.; Hajek, M.; Rauleder, J. Computational flight path analysis of a helicopter in an offshore wind farm using a lattice-boltzmann method. In Proceedings of the AIAA Science and Technology Forum and Exposition (SciTech), Virtual Conference, 19–21 January 2021.
8. Štrbac, A.; Martini, T.; Greiwe, D.H.; Hoffmann, F.; Jones, M. Analysis of Rotorcraft Wind Turbine Wake Encounters using Piloted Simulation. *CEAS Aeronaut. J.* **2021**, *12*, 273–290. [CrossRef]
9. Ramirez, L.; Fraile, D.; Brindley, G. *Offshore Wind in Europe—Key Trends and Statistics 2020*; Technical Report; WindEurope: Brussels, Belgium, 2021.
10. Komusanac, I.; Brindley, G.; Fraile, D.; Ramirez, L. *Wind Energy in Europe—2020 Statistics and the Outlook for 2021–2025*; Technical Report; WindEurope: Brussels, Belgium, 2021.
11. Dalgic, Y.; Lazakis, I.; Turan, O. Investigation of Optimum Crew Transfer Vessel Fleet for Offshore Wind Farm Maintenance Operations. *Wind Eng.* **2015**, *39*, 31–52. [CrossRef]
12. Muller, M.; Greenwood, R.; Richards, M.; Bark, L. *Survey and Analysis of Rotorcraft Flotation Systems*; Technical Report DOT/FAA/AR-95/53; U.S. Department of Transportation (DOT), Federal Aviation Administration (FAA): Washington, DC, USA, 1996.
13. Padfield, G.D.; Manimala, B.; Turner, G.P. A Severity Analysis for Rotorcraft Encounters with Vortex Wakes. *J. Am. Helicopter Soc.* **2004**, *49*, 445–456. [CrossRef]
14. van der Wall, B.G.; Fischenberg, D.; Lehmann, P.H.; van der Wall, L.B. Impact of Wind Energy Rotor Wakes on Fixed-Wing Aircraft and Helicopters. In Proceedings of the 42nd European Rotorcraft Forum, Lille, France, 5–8 September 2016.
15. van der Wall, B.G.; Lehmann, P.H. About the Impact of Wind Turbine Blade Tip Vortices on Helicopter Rotor. *CEAS Aeronaut. J.* **2018**, *9*, 67–84. [CrossRef]
16. van der Wall, B.G. Rotor Thrust and Power Variations during In-Plane and Orthogonal Vortex Interaction. In Proceedings of the 7th Asian/Australian Rotorcraft Forum, Jeju Island, Korea, 30 October–1 November 2018.
17. van der Wall, B.G. Impact of Vortex—Wake Interference on Rotor Trim; In Proceedings of the VFS Transformative Vertical Flight, San Jose, CA, USA, 21–23 January 2020.
18. van der Wall, B.G. Comparison of Different Approaches for Modeling Vortex—Rotor Wake Interference on Rotor Trim. In Proceedings of the 76th Annual Forum and Technology Display, addedOnline, 6–8 October 2020.
19. Bakker, R.; Visingardi, A.; van der Wall, B.G.; Voutsinas, S.; Basset, P.M.; Campagnolo, F.; Pavel, M.; Barakos, G.; White, M. Wind Turbine Wakes and Helicopter Operations—An Overview of the Garteur HC-AG23 Activities. In Proceedings of the 44th European Rotorcraft Forum, Delft, The Netherlands, 18–21 September 2018.
20. Hoffmann, F. *Rechtliche und flugbetriebliche Grundlagen für Hubschraubereinsätze in Offshore-Windparks*; Technical Report DLR-IB-FT-BS-2017-134; Deutsches Zentrum für Luft-und Raumfahrt (DLR): Braunschweig, Germany, 2018.
21. Anon. Bekanntmachung der Gemeinsamen Grundsätze des Bundes und der Länder über Windenbetriebsflächen auf Windenergieanlagen: GGBL-WBF, Bundesministerium für Verkehr, Bau und Stadtentwicklung (BMVBS). Available online: http://www.verwaltungsvorschriften-im-internet.de/bsvwvbund_18012012_LR116116521.htm (accessed on 27 November 2021).
22. Jonkman, J.; Butterfield, S.; Musial, W.; Scott, G. *Definition of a 5 MW Reference Wind Turbine for Offshore System Development*; Technical Report NREL/TP-500-38060; National Renewable Energy Laboratory (NREL): Golden, CO, USA, 2009.
23. Anon. North West Shelf Data Portal, Bundesamt für Seeschifffahrt und Hydrographie (BSH). Available online: http://nwsportal.bsh.de/ (accessed on 27 November 2018).
24. Kroll, N.; Faßbender, J. MEGAFLOW—Numerical Flow Simulation for Aircraft Design. In *MEGAFLOW Symposium*; Springer: Berlin/Heidelberg, Germany, 2005.
25. Jameson, A.; Schmidt, W.; Turkel, E. Numerical solution of the Euler equations by finite volume methods using Runge Kutta time stepping schemes. In Proceedings of the 14th Fluid and Plasma Dynamics Conference, Palo Alto, CA, USA, 23–25 June 1981.
26. Kowarsch, U.; Keßler, M.; Krämer, E. High Order CFD-Simulation of the Rotor-Fuselage Interaction. Available online: https://dspace-erf.nlr.nl/xmlui/bitstream/handle/20.500.11881/605/erf2013_091.pdf?sequence=1 (accessed on 27 November 2021).
27. Klein, L.; Gude, J.; Wenz, F.; Lutz, T.; Krämer, E. Advanced computational fluid dynamics (CFD)–multi-body simulation (MBS) coupling to assess low-frequency emissions from wind turbines. *Wind Energy Sci.* **2018**, *3*, 713–728. [CrossRef]
28. Weihing, P.; Schulz, C.; Lutz, T.; Krämer, E. Comparison of the actuator line model with fully resolved simulations in complex environmental conditions. *J. Phys. Conf. Ser.* **2017**, *854*, 012049. [CrossRef]
29. Troldborg, N.; Zahle, F.; Réthoré, P.E.; Sørensen, N. Comparison of wind turbine wake properties in non-sheared inflow predicted by different CFD rotor models. *Wind Energy* **2014**, *18*, 1239–1250. [CrossRef]
30. Shives, M.; Crawford, C. Mesh and load distribution requirements for actuator line CFD simulations. *Wind Energy* **2013**, *16*, 1183–1196. [CrossRef]
31. Troldborg, N. Actuator Line Modeling of Wind Turbine Wakes. Ph.D. Thesis, Technical University of Denmark, Lyngby, Denmark, June 2008.
32. Türk, M. Ermittlung Designrelevanter Belastungsparameter für Offshore-Windkraftanlagen. Ph.D. Thesis, Universität zu Köln, Köln, Germany, 2008.
33. Mann, J. The spatial structure of neutral atmospheric surface-layer turbulence. *J. Fluid Mech.* **1994**, *273*, 141–168. [CrossRef]

34. Peña, A.; Gryning, S.E.; Mann, J. On the length-scale of the wind profile. *Q. J. R. Meteorol. Soc.* **2010**, *136*, 2119–2131. [CrossRef]
35. Sherry, M.; Nemes, A.; Lo Jacono, D.; Blackburn, H.M.; Sheridan, J. The interaction of helical tip and root vortices in a wind turbine wake. *Phys. Fluids* **2013**, *25*, 117102. [CrossRef]
36. Menter, F.R. Two-equation eddy-viscosity turbulence models for engineering applications. *AIAA J.* **1994**, *32*, 1598–1605. [CrossRef]
37. Rowe, S.J.; Howson, D.; Turner, G. A turbulence criterion for safe helicopter operations to offshore installations. *Aeronaut. J.* **2006**, *110*, 749–758. [CrossRef]
38. Duda, H.; Gerlach, T.; Advani, S.; Potter, M. Design of the DLR AVES Research Flight Simulator. In Proceedings of the AIAA Modeling and Simulation Technologies (MST) Conference, Boston, MA, USA, 21–23 May 2013.
39. Kaletka, J.; Kurscheid, H.; Butter, U. FHS, the New Research Helicopter: Ready for Service. In Proceedings of the 29th European Rotorcraft Forum, Friedrichshafen, Germany, 16–18 September 2003.
40. Maibach, M.J.; Jones, M.; Štrbac, A. Development of a Simulation Environment for Maritime Rotorcraft Research Applications. In Proceedings of the Deutscher Luft- und Raumfahrtkongress (DLRK), Online conference, 3 September 2020.
41. Lehmann, P.H.; Jones, M.; Höfinger, M. Impact of Turbulence and Degraded Visual Environment on Pilot Workload. *CEAS Aeronaut. J.* **2017**, *8*, 413–428. [CrossRef]
42. Hamers, M.; von Grünhagen, W. Nonlinear Helicopter Model Validation Applied to Realtime Simulations. In Proceedings of the 53rd American Helicopter Society Annual Forum, Virginia Beach, VA, USA, 29 April–1 May 1997.
43. Pitt, D.M.; Peters, D.A. Theoretical Prediction of Dynamic-Inflow Derivatives. *Vertica* **1981**, *5*, 21–34.
44. Anon. *Aeronautical Design Standard 33E PRF*; Technical Report ADS-33E-PRF; United States Army Aviation and Missile Command (AMCOM): Redstone, AL, USA, 2000.
45. Anon. *Defence Standard 00-970, Design and Airworthiness Requirements for Service Aircraft, Part 7—Rotorcraft*; Technical Report DEF STAN 00-970 Part 7/2—Section 9; Ministry of Defence: Glasgow, UK, 2007.
46. Cooper, G.E.; Harper, R.P. *The Use of Pilot Rating in the Evaluation of Aircraft Handling Qualities*; Technical Report TND 5153; National Aeronautics and Space Administration (NASA): Washington, DC, USA, 1969.
47. Anon. *Advisory Circular AC 29-2C—Certification of Transport Category Rotorcraft*; Technical Report AC 29-2C; U.S. Department of Transportation (DOT), Federal Aviation Administration (FAA): Washington, DC, USA, 2014.

MDPI
St. Alban-Anlage 66
4052 Basel
Switzerland
Tel. +41 61 683 77 34
Fax +41 61 302 89 18
www.mdpi.com

Energies Editorial Office
E-mail: energies@mdpi.com
www.mdpi.com/journal/energies